ICE AGES

THEIR NATURE AND EFFECTS

ICE AGES

THEIR NATURE
AND EFFECTS

IAN CORNWALL

London
JOHN BAKER LTD

New York
HUMANITIES PRESS INC

© 1970 IAN CORNWALL
First published in 1970 by
JOHN BAKER (PUBLISHERS) LTD
5 Royal Opera Arcade
Pall Mall, London SW1
SBN 212 98375 X

First published in
the U.S.A. in 1970 by
HUMANITIES PRESS INC
303 Park Av South
New York NY 10010

SBN 391 00094 2

Printed in Great Britain at
THE CURWEN PRESS LTD

Contents

List of figures

List of plates

Introduction

Man, as we know him today, is a comparatively new arrival on Earth – probably not more than a quarter of a million years old, as a species. His less highly-developed forebears, from whom he presumably evolved, held the stage for long before that, and, at present, the earliest known Hominid (a member of the Family which includes all the different human species) dates back to perhaps a million and three quarter years before present (B.P.). That point in time falls in the earlier part of the Pleistocene, the last and shortest of the Periods recognized in the succession of events in geological history (Fig. 1). Most of the evidence for this, and for the long and even remoter times before the appearance of man, comes from geology.

ERA	PERIOD	DATES BEFORE PRESENT (B.P.)
QUATERNARY	Holocene	──10,000 years
	Pleistocene	──1.5 to 3 million
TERTIARY	Pliocene	──15 million
	Miocene	──35 million
	Oligocene	──50 million
	Eocene	──86 million
MESOZOIC	Cretaceous	

1 Later part of the geological column, with approximate limiting dates of boundaries.

ICE AGES: THEIR NATURE AND EFFECTS

The geology of the Pleistocene presents problems which are peculiar to it. Over vast areas of the North Temperate zone are found superficial deposits quite different in character from those representing most other Periods and these, for the earlier geologists, were hard to account for by the action of any of the natural agencies with which they were familiar. Thick sheets of 'till' or boulder-clay – an unstratified and ill-sorted material comprising rock-fragments of all sizes, from huge boulders down to impalpably fine particles – mantled whole countrysides. Their origin was made still more mysterious by their often numerous 'erratics' – fragments of rocks obviously foreign to the locality, apparently having been transported for great distances from their nearest known outcrops in the 'solid'. With the boulder-clays could be associated spreads of more or less water-sorted gravels, sands and 'brickearths' – the last a convenient generic term for sandy or silty clays suitable for brick-making. There were, moreover, areas of older rock-outcrops, not covered by anything more recent, which were smoothed and scratched, or broken and contorted, as by some giant power. Most of the agencies of decay, erosion and transport of rock-materials which might, any day, be seen in action, seemed to be much gentler and more gradual and none of them manifestly produced the same effects.

The geologists of the eighteenth and early nineteenth centuries set these most superficial deposits and phenomena apart from the more familiar marine sediments, igneous intrusions or altered rocks such as schists with which they were familiar and which made up nearly all the rest of the known geological materials. They were called 'Drift' – a title which did not in any way pre-judge their mode of formation.

Unable plausibly to explain the Drift, most earlier inquirers had recourse to tradition and free-ranging imagination. Since, from its invariably superficial position, the Drift was clearly assignable to fairly recent geological times, the Flood of Noah, as described in the Old Testament, was summoned up to account for it. Vast cataclysmic 'waves of translation' were postulated, which could pick up and transport at once house-sized boulders and tiny clay-particles, for many miles over hill and dale. Indeed, in the case of some erratic rocks of obviously Scandinavian origin found in eastern England, it was necessary to suppose that they had thus been conveyed right across the North Sea. So the Drift became also the 'Diluvium', product of the Deluge, a unique event which divided the geological Present – the period of historical mankind – from the remoter Past – Antediluvian times.

To a few this attribution remained unsatisfactory, because of the many obvious objections to such a theory raised by any objective examination of the Drift deposits.

Even granted 'waves of translation' on a scale never witnessed today, any body of water, in their experience, would have sorted and stratified its sediments more distinctly as they were being laid down. Again, though, for instance, pebbles may manifestly be formed from angular rock-fragments by the action of waves and currents, they are never, in the process, scored and scratched like the stones in the Drift, nor is any solid rock-floor, over which they may be transported by water, striated, as is that underlying the Drift.

In the first third of the nineteenth century, continental geologists, among them Schimper, Venetz and Charpentier, studied the deposits and effects of modern Alpine glaciers and recognized their close resemblance to those of the Drifts of northern Europe, though the latter were on an immeasurably larger scale. It was therefore concluded that the Drifts must be attributed to the action of an ice sheet, marine in origin, centred on the North Pole and thence spreading southwards in all directions. Joshua Trimmer found marine shells in the Drift at a height of 1,350 feet on Moel Tryfaen, in North Wales and this, and other similar cases, made it appear that the advance of the hypothetical polar ice had been accompanied by a marine submergence of the land to perhaps 1500 feet. A sea was visualized, covered with floe-ice and bergs, and these would have rafted stones and shells and clay together, from which the boulder-clay had been deposited on the sea-floor at their melting.

Dean William Buckland, Doctor of Divinity and the first Reader in Geology at the University of Oxford, had published in 1823 his *Reliquiae Diluvianae*, a treatise on cave-deposits, 'diluvial' gravels and their contained fossils, 'attesting the Action of a Universal Deluge'. Soon afterwards he met Louis Agassiz, the great Swiss palaeontologist and glaciologist, who, though at first himself sceptical, had finally become convinced, following Schimper and Venetz, that the Alpine Drift was the product of glaciers, not of the sea or of any Deluge – indeed of a sheet of land-ice originating in the Alps but vastly more extensive than the modern valley-glaciers. On the ground, Agassiz showed Buckland the evidences for this conclusion and the traces of former ice-sheets extending as far as Lyon. Buckland, too, was convinced and, at his invitation, Agassiz later came to Britain and here, also, recognized all the features of widespread glacial action, in Scotland and southwards over much of England.

Sir Charles Lyell, a pupil of Buckland, in his *Principles of Geology* (1830–3) adopted the doctrine of Uniform Causes, first enunciated by James Hutton in his *Theory of the Earth, with Proofs and Illustrations* (1795), and this doctrine, with only minor reservations, continues to hold the field. It is maintained that only agencies and processes still to be seen today as effective are admissible to explain geological phenomena of the past. Lyell's *Principles* is the first full generalization of geological knowledge to its date and remains the foundation-stone of modern geology. Catastrophism, in the sense of the unrestricted speculations of the old Cosmogonists, has ever since been discredited. Though there are still plenty of grounds for disagreement and controversy over questions of Pleistocene geology, the glacial origin of the Drifts in middle latitudes is fully established today.

The classic synthesis of the evidence for Pleistocene glaciation is James Geikie's *The Great Ice Age* (1874, 3rd ed. 1894), still eminently readable and instructive today. The work is based on the example of the detailed field-evidence in Scotland, but proceeds thence to describe what was known at the time in Europe as a whole, in Asia, Africa, Australia and New Zealand, and, in a chapter contributed by T. C. Chamberlin, of North America. Geikie details the evidence for a subdivision of the Ice Age into six glacial epochs, separated by five interglacials, all preceding the establishment of the present day (Postglacial) climatic conditions. The last three of his epochs are nowadays assigned to 'stadia' – lesser advance-phases – of the fourth and last of Penck & Brückner's (1909) main glaciations, named from the north Alpine river-valleys, Günz, Mindel, Riss and Würm, in which their deposits are typically exposed.

The glacial subdivisions of Penck & Brückner were based on the superposition of the moraines and their connections with different spreads of outwash gravels and *Deckenschotter*, climatic river-terraces in the north Alpine area. Interglacials were evidenced by the weathered surfaces on the moraines and occasionally by deposits indicating temperate intervals by their contained fossils; the duration of the Inter-glacials by the depths to which such chemical alteration had penetrated into each series. Properly speaking, these findings, and the names applied to the different phases, are applicable only in the area where they were originally studied. Later work, however, has shown that they have correlatives elsewhere, and, though local names have often been given to these, their correspondence with the Alpine sequence has often been pretty firmly established. Thus, Elster, Saale and Warthe/Weichsel seem in north Germany to correspond with Mindel, Riss and Würm (Günz, so far as is known at present, being unrepresented there). In south-eastern England, the Older Drift (corresponding to Alpine Mindel and Riss) boulder-clays have lately been given the local names of Lowestoft and Gipping; the Younger Drifts, spreading southwards at their greatest extent to Lincolnshire and north Norfolk from Scotland, correspond to the several subdivisions of the Weichsel of north Germany.

A detailed general subdivision and correlation of European continental glacial and interglacial deposits was proposed by W. Soergel (1925). This showed two phases each for the first three main glaciations and three for the last, these 'Stadia' being interrupted by minor retreat-phases called Interstadials.

Twenty years later, the whole subject was summarized and interpreted in the light of research to its date by Soergel's one-time colleague and follower, F. E. Zeuner (1945), with special reference to the chronology. In the more than twenty years which have since elapsed, the already vast and multilingual literature of Pleistocene studies has again grown enormously. Relevant additions to knowledge have come from climatology, soil-science, oceanography, hydrography and, especially in the field of chronology, nuclear physics – to name but a few of the contributing scientific disciplines.

In addition to their value, as yielding background information and a dated framework of natural historical events for the student of early man (and, indeed, for all other biological and ecological sciences, for which they provide the time-dimension) the Pleistocene Ice-Ages have an intrinsic interest, which requires no excuse for its pursuit.

CHAPTER 1

Physical effects of glaciers and land-ice

Glaciers and ice-sheets tend to develop on land when local seasonal temperatures are predominantly below freezing-point, so that the rate of accumulation of snow exceeds that of wastage by melting and evaporation. Mere low temperatures, however, are not sufficient, for, without accompanying high precipitation, predominantly in the form of snow, accumulation cannot take place. The snow-line, down to sea-level in Polar regions, lies between 17,000 and 18,000 ft. (5,600 to 5,930 m.) at the Equator and is a little above 5,000 ft. (1,650 m.) in the latitude of Britain.

Formation of land-ice thus begins in mountainous areas, generally in fairly high latitudes, where both these conditions are satisfied.

Even in such high snowfields, daily and seasonal ambient temperatures may frequently fluctuate about the freezing-point of water. Thus, in time, alternate softening and re-freezing of freshly fallen, loose, powdery snow will result in gradual compaction under its own weight and expulsion of air trapped among the original ice-crystals, with recrystallization and the formation of denser, coherent 'old' snow (*firn* in German, *névé* in French).

The process continues, with the added weight above it of later snow-falls, until the still relatively porous firn is converted into solid, finally almost glassy-clear, ice. The individual interlocking crystals of solid, pure ice are separated from each other by exceedingly thin films of liquid water containing dissolved impurities (salts, etc.) which lower their freezing-point, and it is along these planes of separation that the crystals are able to glide upon each other when subjected to differential forces, so that ice behaves like a very viscous fluid. On a relatively steep slope, therefore, ice will be deformed by the force of gravity and slowly flow downhill.

The resulting ice-stream, or *glacier*, occupies the lowest ground presented to its flow – the bottom of a valley sloping from the high snow-fields to lower levels. As its elevation decreases, so does the average ambient temperature rise, until, at a certain point, seasonal thawing outpaces the rate of advance of the glacier-front and the ice gives rise to a stream of *meltwater*.

Not only are there seasonal variations in this equilibrium between ice-accumulation and melting, and so in the position of the ice-front, but longer-term climatic changes may cause the glacier to advance on the whole over a period of years and again to retreat when, for some time, the rate of wastage tends to surpass that of growth.

A growing valley-glacier may entirely fill its valley and overflow, to merge with the ice of adjacent valleys. An *ice-cap* is thus formed, completely mantling at least the higher parts of the mountain massif, and, if its increase in volume is sufficient, and the local snow-line low enough, the cap may become a widespread *ice-sheet*, covering foothills and lower ground, beyond the mountains in which it originates – *piedmont-glacier*. Ice-sheets of this kind exist today in Antarctica and Greenland. Geological and topographical evidence shows that, at different times in the past, notably in the Pleistocene and in North America and Europe, similar, but even more vast, sheets of *land-ice* were widely extended.

EROSIONAL FEATURES OF GLACIATION. GLACIAL DEPOSITS

A glacier, originating in high snow-fields and moving down its valley towards its termination at the point where wasting by melting exceeds the rate of advance, gives rise to characteristic features of erosion of the bedrock and deposition of sediments, and these may persist for long after changed conditions have caused the ice to disappear entirely.

At the head-wall of the snow-field (Fig. 2), firn and the underlying ice are frozen to the rock and are under tension as the body of the glacier moves away downhill. Fragments of the head-wall are thus torn away or 'plucked' by the ice from the bedrock, causing the wall to recede, as it is eroded, in a rough semi-circle. The combined erosion of the head-wall and the pull away from it of the ice form a deep crevasse, or *bergschrund*, at the head of the firn-field. Where this is intermittently open above, it receives angular, frost-weathered rock-fragments and meltwater from the wall, which eventually reach the sole of the ice-field. The bergschrund is periodically re-filled with fresh snow and avalanching from the head-wall, to continue the process.

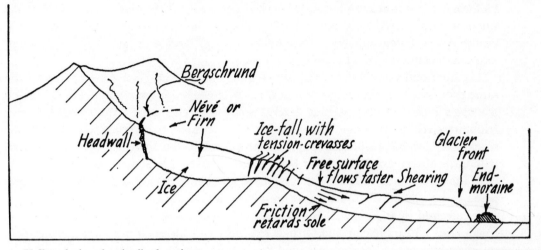

2. Valley glacier: longitudinal section.

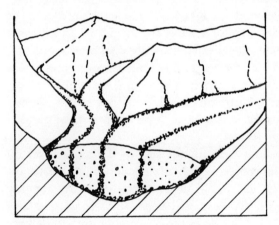

3. Confluent valley-glaciers, with lateral, medial, englacial and ground-moraines (transverse).

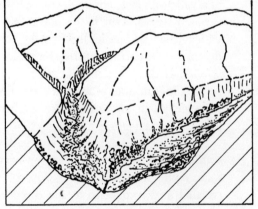

4. The same valley system deglaciated: characteristic U-shape, hanging tributaries, waterfalls, scree-slopes etc.

The floor of the ice-field is hollowed out, ground and abraded, not only by the mass of the slowly moving ice itself, but by the 'teeth' provided by the rock-fragments embedded in the sole of the glacier. On the disappearance of the ice-field, its former area of accumulation is marked by the semi-circular head-wall and the smoothed and scored rocks of the basin near to its foot, forming a characteristic and unmistakable topographical feature called a *cwm*, *corrie* or *cirque*. The now-vacant basin may contain a small lake or *lochan* (Scots).

The floor and sides of the valley which once contained a glacier are carved by the same processes to a characteristically smooth U-shaped cross-section (Fig. 4) which truncates most formerly-existing lateral spurs and cuts across the valley-floors of tributary streams. Where a valley-glacier has retreated or completely vanished, the transverse U-shape, truncated spurs and lateral valleys left 'hanging' (so that modern tributaries often reach the main stream by waterfalls) clearly point to former ice-action.

In the case of mountain ice-caps or inland ice-sheets, the entire former topography tends to be planed off by the passage of the ice. Only emergent peaks and crags never covered by it (*nunataks*), but bypassed by the ice on either side, still show the rough cliffs and pinnacles due to their exposure to continued frost-weathering, now standing high above the ice-smoothed lower topography.

Rock-waste, thus weathered by frost-and-thaw off high peaks and bordering sides of a valley-glacier, comes eventually to cover the edges of the glacier at the foot of the exposed slopes. This is the *lateral moraine*. Where a tributary glacier in an adjacent valley joins the main ice-stream, the two nearer lateral moraines meet and are, downstream, carried on by the main ice at some distance from its edges as a *medial moraine*. If there are several such tributary glaciers in a system, there will be a corresponding number of medial moraines downstream of their junctions, so that, from a distance, as in an aerial view, such a glacier is seen to have a number of parallel longitudinal stripes of morainic débris on its surface (Fig. 4).

Any large inequalities in the valley-floor beneath the sole of the ice will set up sudden local forces of tension and compression which the glacier is unable to absorb, owing to the high viscosity of the ice. Series of fractures, generally roughly parallel and transverse to the direction of ice-flow, will open up as *crevasses* at any abrupt increase in gradient, or, if the downhill side of the obstruction should be very steep, even form a tumultuous *ice-fall*. When the slope eases, beyond the obstruction, the crevasses will once more close.

Moraine material lying on the surface of a glacier will be entrapped by gravity or washed by meltwater-streams down the open crevasses and so perhaps directly reach the sole of the ice, to join the rock materials already plucked and abraded from the valley floor. This mass of débris, beneath, and within, the lower part of the glacier forms the *ground moraine*. Not all the morainic waste falling into crevasses will reach the base of the ice. Penetrating only part of the way and caught by the re-closing of the crevasse a little further on, this will be embedded in the mass of the ice and is called *englacial moraine*.

At its terminus, where the glacier melts in summer as fast as it is pressed forward by the weight of ice behind it, all this transported rock-waste is deposited, if there is space, as a more or less crescent-shaped *terminal moraine*, consisting of haphazardly-dumped materials of all sizes and shapes. An advance from any cause will plough up, re-work and redeposit any former terminal moraines. A renewed standstill will allow a fresh accumulation to begin at the nose of the glacier. A retreat will result in abandonment of the terminal moraine, marking a previous standstill-position of the ice-front, and cause the beginning of a fresh morainic accumulation at the existing terminus of the glacier.

Ice-caps or -sheets of larger extent than valley-glaciers, having no emergent nunataks, have no source of surface morainic material. They do, of course, plough up and incorporate to some extent in their thickness, fragments (and even huge masses) of the rocks over which they have passed, and these they deposit at their wasting fronts as terminal moraines, marking their former maximum advances. Other terminal moraines, of lesser advances or of stages of standstill during the latest retreat, may be laid down within them.

Moraine material consists of a mixture of all the different sorts of rocks traversed by a glacier or ice-sheet, of all sizes, from the vast blocks mentioned above, through boulders to gravel, sand, silt and clay – the last being the finest rock-flour, ground to impalpable powder by the weight and abrasive power of the ice moving over standing rock-surfaces. This mixture is called *boulder-clay* or *till*. Boulder-clays of pre-Pleistocene glaciations, hundreds of millions of years old and now converted into hard stone, are termed *tillites*.

Outcrops of hard rocks *in situ*, which have been abraded, rounded and striated by a recent cover of glacier-ice, present a peculiar appearance in a landscape and are called *roches moutonnées*, from their resemblance to groups of grazing or resting sheep when viewed from some little distance. Vegetation tends to be confined to the moister hollows, where soil-material collects between the convex 'sheep'-backs. The smooth, rounded slight eminences are scoured by wind-driven rain and remain relatively bare. The frequently pale grey colour of their weathered surfaces adds to the resemblance (Fig. 5).

5. Roches moutonnées. *Sutherland, Scotland.*

Just as the underlying rocks of an ice-mass are polished and striated by the passage of the glacier entraining rock-waste in its sole, so do the transported blocks become battered and scratched, smoothed and rounded during their journey – in the case of some far-travelled examples, often for many hundreds of miles. Such glacial boulders and pebbles, if of chemically-resistant materials such as quartz or sandstones, may readily be recognized thousands of years since their deposition, wherever they may be found. On the Yorkshire Moors, which were never overflowed by the Last Glaciation (Weichsel) ice-sheet, many such foreign pebbles are still to be found, and can only be attributed to a still earlier glaciation which must have overridden the hills, all other traces of which have since been lost by erosion and weathering. Some will come to lie very far from their places of origin, among others derived from no great distance and many more composed of the local rock over which the ice was travelling at the time of their deposition. If not belonging to the 'solid' on which they now lie, such 'foreigners' are known as *erratic blocks* (Fig. 6), or simply *erratics*, and their nature may tell the geologist exactly whence and by what route the ice came which transported them.

Some such ice-transported boulders, often, but not necessarily, erratics, may have been gently deposited by slowly melting ice in the most unlikely and precarious

6. *A perched erratic block of (older) Silurian grit on a pavement of Carboniferous limestone. Yorkshire.*

positions, propped hazardously on steep slopes, as if by a playful giant hand. These are *perched blocks* and, among the other features described here, are a sure indication of comparatively recent ice-action, for the ordinary processes of denudation will inevitably, inside a few thousand years at most, unseat them and topple them down-hill to some lie of less unstable equilibrium.

Ice-scratches (*striae*) (Fig. 7), inflicted on glaciated *in situ* rock-surfaces will betray the general direction of movement of the ice which once crossed them. So do the more frequent orientations of the long axes of pebbles in ground-moraine show it, when no hard underlying rocks, capable of preserving striae, are in evidence.

Where an ice-sheet has recently covered relatively flat terrain, it has often left some fairly large-scale topographical features which, for a time at least, survive its disappearance (Fig. 8).

Drumlins are low hillocks of boulder-clay, seldom more than 100 feet (30 m.) in height, of characteristically rounded outline, generally of a more or less oval figure in plan. They are frequently closely grouped, often in somewhat ordered ranks, échelons or columns, their longer axes strictly parallel to the direction of ice-movement (Fig. 9).

Rock-basins may have been locally excavated on faults, joints and other longi-tudinal zones of weakness in the bedrock and these may, today, hold lakes having a

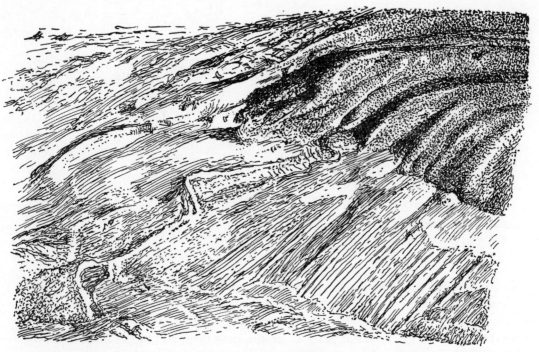

7. A retreating glacier front (top right) *exposes smoothed and striated rock-surfaces and an esker marking the former course of a sub-glacial stream. Woodworth Glacier, Alaska.*

main elongation in the direction of ice-flow. On glaciated coasts, such basins may form *fjords*, or *sea-lochs*, if subsequently flooded by salt water. Other lakes or tarns may be *kettle-holes* (Fig. 10), depressions in morainic débris once occupied by isolated masses of *dead ice*, detached outliers of the main glacier, cut off by the retreat and buried in moraine. They survive only briefly the final retreat of the glacier-front and when they melt leave often undrained depressions.

Lateral channels may be cut by draining meltwater along the edge of a melting ice-sheet, partly in the waning ice and partly in the solid underlying rock. They remain, sometimes high on a hillside, completely out of context in the ice-free topography. In places such streams of ice-water may be temporarily dammed back, during the general retreat, by some surviving tongue of ice or the remains of a tributary valley-glacier. In that case they form *ice-dammed lakes* (Fig. 11), the overflow from which

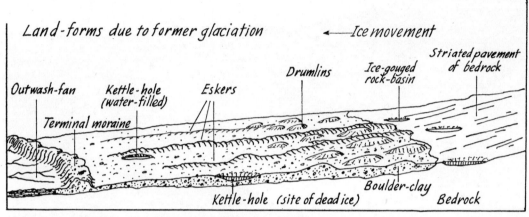

8. *Diagram of glacial topographical features after retreat of ice.*

9. *Drumlins. Lanarkshire, Scotland.*

may cut a *gorge* or *overflow-channel*, in solid rock over the lowest point of a divide (Fig. 12).

Part of the drainage from a melting ice-edge is by *sub-glacial* or *englacial streams* flowing in longitudinal cracks or ice-tunnels melted and eroded in the very body of the ice-sheet. Such streams gather meltwater from far up the glacier, and this finds its way down crevasses. When flowing strongly, the water is charged with rock-débris and sediments derived from the ice. The coarser part of the load is readily dropped whenever their current slackens and builds up a bed of somewhat washed, sorted and stratified gravel, which is retained by the ice-walls confining the flow. On final

25

10. Kettle hole, waterfilled. Orkney.

melting of the ice, this lateral support is removed and the gravels slump transversely to the direction of former flow. This forms an irregular ridge, more or less continuous and meandering, with sides at the angle of rest of the material and with little remaining stratification – an *esker* (Figs. 8, 13). *Kames* are seen today as more or less isolated mounds of similar material, once laid down as the delta or fan of gravel at the mouth of an englacial stream. Like an esker, it will have slumped on removal of support by the ice at its back and margins.

OUTWASH DEPOSITS

Meltwater flowing away from the ice-front forms *braided streams*, which are continually altering course, branching and rejoining one another, as they transport, sort and re-deposit an overload of *glaci-fluvial outwash-gravels* and *-sands* (Fig. 14). Such outwash-plains, or *sandrs*, often extend for many miles beyond the ice-front, and here they are exposed to cold, dry winds blowing off the ice-sheet. Much fine dust from the bare drying surface may be picked up by the wind and blown away,

11. Märjelen See – a lake of meltwater dammed up by the ice of the Aletsch Glacier. Switzerland.

12. Overflow channel, now dry, of a former ice-dammed lake. Langdale Gorge, North Riding, Yorkshire.

27

13. Esker, Finland. The lake floods an irregular outwash-plain.

leaving a *pavement* of coarser materials from the interstices and surfaces of which much of the fines has been *deflated*. This wind-borne dust may be transported for hundreds of miles from its place of origin and be re-deposited far away as *loess* (see below, p. 32).

Further downstream, the outwash material, now becoming somewhat better sorted in particle size, is rearranged into an only gently sloping surface, the *floodplain* (Fig. 15) of the stream-valley. This consists of truly fluviatile stratified deposits of gravels, sands and floodloams. Such deposits still contain the erratics and scarcely yet rounded and abraded larger stones, many of which will still betray their former glacial origin.

14. Braided streams of glacial meltwater crossing outwash-plain. Headwaters of Rangitati River, South Island, New Zealand.

Under a cold or cool climate, with periodic frosts at least in winter, the sides of a river-valley contribute a considerable proportion of the floodplain sediments, in the form of frost-weathered, more or less angular, rock-fragments, tumbling, sludging and washing down the slopes through the influence of denudation under gravity. The movement of these materials is aided by intermittent freezing and thawing of the surface with perhaps still at least seasonally frozen subsoil. Such ill-sorted *solifluction*-deposits are likely to become interstratified with stream-borne materials towards the margins of the floodplain, to the extent that, on reaching it, they are not completely washed and re-sorted by the action of the river. The water-filled channel of the stream meanders back and forth over the floodplain, in part re-working its own earlier deposits, but, as long as the supply of rock-waste from the valley-sides exceeds the river's power to transport it further down its course, continually adding to the already accumulated débris filling the valley and building up

15. River floodplain with meanders.

the floodplain-surface. Excluding other factors affecting a river's performance, therefore, continued supply of glacial and periglacial materials under a cold climate favours this *climatic aggradation* of the valley.

PERIGLACIAL ZONE

This far down the course of a river draining a glacier or ice-sheet, we are well outside the contemporary glacial environment – in the *periglacial zone*. Owing to the relative proximity of the ice, this zone may, nevertheless, be subject to climatic influences directly due to its presence, and exhibit geological phenomena clearly emblematic of those conditions.

In the case of local ice-caps or valley-glaciers, at the most a few tens of miles in length or diameter, the width of the periglacial zone affected will be slight – perhaps only a few hundred metres down-slope from the ice-edge, depending on the situation of the place in latitude (i.e. its maximum sea-level average temperature) and the steepness or otherwise of the consequent temperature-gradient with decreasing altitude. Where there was a vast mass of land-ice, on the other hand, as over northern Europe or North America in the Pleistocene, some, or even many, hundreds of

kilometres in diameter, its periglacial effect extended a very long way from
front. This was mainly due to the formation of a persistent *glacial anticyclone*
ice-covered area larger than a few hundreds of kilometres in diameter. The an
mass in contact with the ice was extremely cold and dense, setting up a correspond-
ingly large area of high atmospheric pressure, such as exists at the Poles today, and
from this dry frost-winds would blow radially outwards, being diverted in a clockwise
sense, as are all moving bodies on the Earth's surface in the northern hemisphere, by
the Coriolis force, reaching the periglacial zone from a north-easterly direction. The
clear skies associated with such a static high-pressure system, moreover, would
favour low night temperatures by loss of heat through radiation to space, so that,
during much of the year, the ground of the periglacial zone was subject to intense
frost. The parts nearer to the ice-front, at least, might not thaw to any depth even
in summer, giving rise to *frozen ground*, *permafrost* or *tjaele*. Further from the ice, the
ground would freeze in depth only in winter.

STRUCTURE-SOILS

The geological signs of former permafrost effects may be seen in fossil *frost-cracks*,
ice-wedges, *frost-polygons* and *reticulated* and *striped formations* of loose stones on ancient
periglacial land-surfaces. Under extremely low temperatures, the ground shrinks
and cracks are formed in roughly polygonal patterns, these varying from a half to
several metres in diameter. Seasonal melting enables surface-water carrying sediments to fill such cracks and the filling, on re-freezing, expands and gradually
widens them with the passing years. Often a still-frozen subsoil will obstruct drainage
by percolation, so that the process is readily repeated.

A seasonally waterlogged, boggy surface is characteristic of areas with permanently-
frozen subsoil. On quite a small scale, alternate freezing and thawing of this wet
superficial layer may set up local 'cells' on level ground, whereby loose stones and
larger solid bodies are gradually brought to the surface centrally by *frost-heaving*.
Water freezes preferentially under a solid stone, lifting it by ice-crystal growth in the
direction of least resistance (i.e. vertically, towards the free surface). On thawing of
the surroundings, finer particles of rock silt down into the spaces occupied by ice-
crystals and prop up the stone in a position higher than before. In this way, coarser
material eventually works to the surface, where, by continued frost-heaving of other

stones behind it, each tends to topple outwards, towards the periphery of the 'cell', on thawing. Eventually the stones come to rest in contact with their opponents from adjacent 'cells', forming more or less regular stone-polygons or continuous networks (*Steinnetze*). On a gentle slope, polygons tend to form more irregular figures, somewhat elongated down-slope, while, on a fairly steep slope, the continuous downhill migration with every surface melting of the polygons gives rise to *stone stripes* running parallel with the steepest gradient. In vertical section, such fossil *structure-soils* may show a series of dis-continuous 'pockets', or *Brodel* in which many stones stand with their longer axes vertical, whereas, in water-laid gravels, they lie predominantly horizontally (Fig. 16).

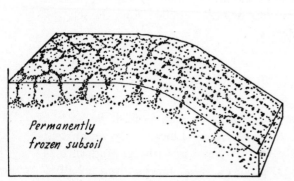 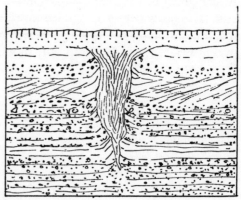

16. Stone polygons and stone stripes (left), *fossil ice-wedge* (right).

LOESS

Dust originating on outwash-plains in the immediate vicinity of an ice-front provides the parent material for important masses of aeolian sediments forming elsewhere.

The fragmentary materials deposited by outwash-streams are largely angular and unsorted as to particle-size. Because of cold winds radiating from the glacial anticyclone, such outwash-plains are much subject to frost action. Exposed rock particles both absorb moisture between their constituent mineral grains and retain it in cracks which they may have suffered in the course of transport by ice. The water freezes overnight, or whenever the temperature falls low enough. Since ice is less dense than liquid water, freezing is accompanied by expansion (by about 10 per cent. of the original volume) so that the cracks and interstices are widened. On thawing, the water produced penetrates still deeper into the widened crack and, filling it completely, once more wedges the walls apart when re-frozen. Thus, frost and thaw, in the presence of some moisture, in time result in comminution of all rock-masses. The process is, however, limited by the size of the particles concerned. Most particles of any rock smaller than about 0·1 mm. are single mineral crystals which, in many cases do not present cracks admitting water in appreciable quantity. Further reduction in size is confined to fractures caused by sudden changes in temperature to which the particle may respond by uneven expansion or contraction, setting up internal stresses. Beyond a minimum particle size of about 2 micrometres (0·002 mm.) even this comminution must cease, for the elasticity of the material is sufficient to enable it to yield to the stress imposed without fracture. Frost weathering thus causes the ultimate formation of quantities of silt-grade (0·06–0·002 mm.) material, which it cannot further reduce to clay (less than 0·002 mm.).

As the cold wind off the ice warms up over the ice-free periglacial area, its relative humidity falls steeply and so sandrs and outwash-fans dry out rapidly as soon as the meltwater drains or seeps away underground. Even a quite moderate wind is able to pick up and transport silt-grade rock-dust, bodily suspended in the air. Only when the wind-velocity is checked does the fine dust settle, perhaps hundreds of kilometres from its point of origin. Vegetation, largely annual grasses, but with some scrubby woody plants in moister places, covered enormous areas of the Pleistocene periglacial zone in the central and western parts of Europe. Near the ground-surface grasses and shrubs checked the dust-laden winds, causing them to deposit part of their load of airborne silt. Protected by the grass-stems and, in due course fixed by their root-mat, the dust-grains were in great part prevented from being picked up again, even by strong winds. Thus, huge accumulations of wind-sorted aeolian sediments, often several metres in thickness, were formed during glacial periods in these *loess-steppes*.

Loess is characterized by its marked sorting, predominantly (often as to 80 per cent. by weight) in the silt grades. Coarser particles (sand) and finer (clay) seldom exceed 20 per cent. together in fresh, unweathered loess. The former, even if wind-transported, must be of more local origin, for grains of sand-size move when driven by wind, in a succession of more or less long leaps, settling again as each strong gust

33

dies away. Clay, on the other hand, is so fine as to remain air-suspended for weeks and months, so that it tends to be carried right away from the source of the material and to lose its identity among other sediments in the ocean basins, which cover some 70 per cent. of the Earth's surface.

Fresh loess is nearly always calcareous, for glacial outwash consists of a mixture of all rocks traversed by the ice and some of these, in any considerable area, are likely to be limestones.

Certainly, not all wind-transported dust is loess. Hot deserts, too, yield plentiful dust to strong winds, as do volcanic explosions in less arid regions, but the constancy of glacial winds, the large areas available for drying outwash deposits not yet fixed by vegetation favour the massive accumulations which we find originating in this particular way, and so are, in the present temperate zone, confidently to be cor-related with glacial climates of the past.

SEA-LEVEL CHANGES

The growth of a vast ice-sheet on land has two other effects. First, the water falling as snow and feeding the ice comes, ultimately, from the oceans. When large areas of the northern continents were mantled in ice, often to a depth of 1,000 m. or more, the amount of water thus immobilized was clearly sufficient appreciably to have reduced the general level of the oceans (*eustatic effect*).

Secondly, the mere dead weight of this mass of ice, imposed for tens of thousands of years at a time on the land-masses was sufficient to cause the loaded continental blocks to sink to some extent into the viscous underlying *mantle*, a deformable layer lying at a depth of some tens of kilometres below the surface (*isostatic effect*).

While this latter is confined to the ice-loaded regions, the former effect is obviously world-wide, so that evidences of former sea-levels, lower or higher than at present, on any stable coastline in the world may confidently be assigned to periods of glaciation and to interglacials respectively. Since the Earth's crust (apart from isostatic de-pression by ice-loading and gradual recovery when the ice-sheet melts) is, however, not everywhere stable today, nor can always be proved to have been so throughout the Pleistocene, this proviso of stability must be reasonably well assured before sea-level evidence anywhere can be usefully interpreted.

The amount of lowering of sea-level due to widespread maximum glaciation of the northern continents may have been of the order of 200 m. (660 ft.). This would have laid bare the submarine continental shelves over the whole world and so have had important effects on climate in particular areas lying near large expanses of dry sea-bed. It would, for example, greatly increase the continentality (seasonal extremes of temperature, decreased rainfall) of lands in north-western Europe at present enjoying relatively equable, moist oceanic conditions. This, alone, would account for the evidence of winter-frozen soils in south Britain and the rest of lowland western Europe, with reindeer ranging as far south as the Pyrenees. Great Britain would then cease to be an island, with much of the North Sea and English Channel areas dry land, and even Ireland would be joined to the rest by landbridges across the present Irish Sea.

Many similar shallow seas the world over would disappear and continental islands be joined for a while to their mainlands. Even in much deeper seas, like the Mediterranean, the fringing shelves would be more or less exposed, so that onshore winds could drive sand-dunes inland from the new coasts, where now there is no extensive foreshore to provide the loose source-material. Ancient aeolian sand accumulations do, in fact, exist near Mediterranean coasts, where none could conceivably be formed at present (Map, Fig. 17).

Since the coastal features of periods of low sea-level are, today, far below mean tide level, they are little known, but evidence of the effects of low seas on the rivers draining into them is perfectly accessible.

A falling sea-level at a river's mouth has the effect of *rejuvenating* the stream, the increased gradient available renewing its erosive power by creating rapids near the mouth. The zone of rapids gradually cuts back upstream by eroding the old bed to a deeper channel. Above the rapids, the longitudinal profile is unchanged, but at the point where renewed erosion begins there will be a sharp increase in rate of fall, showing a 'knick', or downward break in gradient in the smooth curve presented by the former profile. This *knickpoint* marks the position of the head of active erosion and, by cutting away the former bed and floodplain deposits, steadily works its way upstream, leaving below it a new profile, graded to the lower sea-level (Fig. 18).

Provided that the new conditions persist for long enough, the recession of the knickpoint will carry it far enough up the river's course to exempt it from being later involved in any renewed rise in sea-level. Theoretically, it will, thereafter, cut away the older longitudinal profile right back to the river's source. In fact, this increasingly slow process is seldom completed, for, on the time-scale of relatively rapid

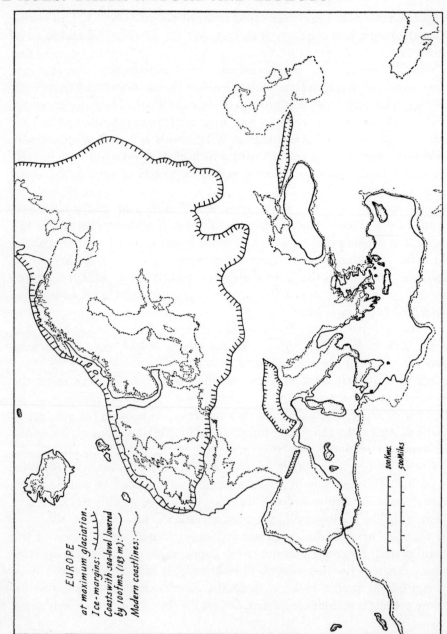

EUROPE
at maximum glaciation.
Ice-margins:
Coasts with sea-level lowered
by 100 fms. (183 m):
Modern coastlines:

17. *Map of maximum glaciation and lowered eustatic sea-level in Europe.*

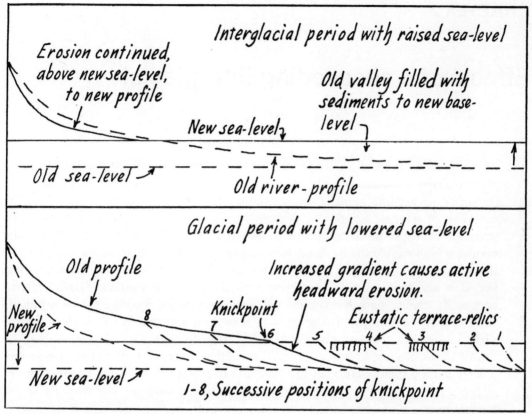

18. *Eustatic rejuvenation of river.*

Pleistocene climatic and consequential changes, it may still be observed in action today. There may, thus, be a succession of knickpoints to be found in a river's profile, each corresponding to a distinct event of rejuvenation which, in the absence of demonstrable tectonic upheaval of the land which it drains, may be attributed each to a former period of low sea-level, corresponding with a glaciation of some importance.

CHAPTER 2

Effects of a succeeding Interglacial

Maximum extension of an ice-sheet into lower latitudes evidently results in more rapid melting and wastage near its margin. The glacial anticyclone established over its centre of accumulation, though maintaining the low temperatures there by exclusion of warmer air-masses, eventually, by its own growth, diverts most depressions from the more central area, and it is these, bringing moisture from the oceans, which enable the ice-sheet to maintain itself, if not to grow further. Reduction of accretion by fending off the depressions means that snow accumulation must first slow and then cease to keep pace with the marginal wastage. Glacial retreat necessarily ensues, though complete disappearance of the ice may not be attained: this probably requires the bringing in of some new external factor, such as a local slight increase in average summer temperatures. Such a slight influence may be caused by an increase in summer radiation received over the centre of accumulation.

Even such a small change in the direction of increased radiation received might trigger a full glacial retreat if it should coincide with an already decreasing access of fresh snow.

Whatever the immediate causes, glacial retreat will begin and continue as soon as overall summer melting outpaces winter snowfall. Not only actual ice-fronts will be affected; the level of the snowline will respond over the whole area of the ice, accelerating melting wherever it surpasses the ice-surface even during only a small part of every year. Towards the extremity and on either side of projecting ice-lobes the rise in ambient temperature will be much more marked than over the ice-mass itself. Ice is not only a very poor conductor of heat but, until it actually melts away it will keep everything with which it is immediately in contact down at the freezing-point. A raised snowline means that, even if precipitation below it is seasonally received as snow, this will not ever persist through the ensuing summer, and summer precipitation will almost invariably be in the form of rain. Only the now reduced area of snowfield above the snowline will continue to contribute to the maintenance of the ice.

Just as in the case of the ice itself, the glacial 'high' will not immediately dissipate: that will be delayed until the ice-covered area has shrunk to some critical size, probably of the order of a few hundreds of kilometres in diameter. Air in contact with it will still be cooled at least to 0°C and its white and shining surface will still continue to reflect into space up to 70 per cent., even, of the possibly increased radiation now falling on it (*albedo* effect), instead of absorbing it as heat.

The rate of melting at the ice-edge, once accelerated, will nevertheless continue of its own momentum. The ground nearby, freed of its cover of ice and snow, will absorb more heat and, because rock has a much lower specific heat than water, will warm up faster and readily yield up such warmth to passing air-streams, which may, in their turn, transfer it to the ice.

GLACIAL LAKES

The retreat of the ice, once well under way, will produce floods of meltwater, as the once pent-up oceans rush down their watercourses to rejoin the sea.

Here and there, in rock-basins hollowed beneath the sole of the ice-streams, or in widening fissures carrying drainage at the margins of valley-glaciers, large volumes of ice-water will accumulate behind any barrier, whether of still-unmelted ice or a permanent topographical feature, and these will overflow their barriers at the lowest point. Waterfalls in these places will cut away fiercely at the obstruction, eroding deep gorges along ice-margins and notching relatively high cols and moraines where the concentrated drainage will break through and tumble down the new-found gradient.

Such glacial lakes are only temporary features but, while they last, their waters may erode enduring features into the topography, such as the famous Parallel Roads of Glenroy (Fig. 19), which are successive wave-cut beaches of an ice-dammed lake. Each is the mark of a standstill in level while the overflow-stream worried away at its contemporary rocky or icy barrier. The retreating ice from time to time laid bare a still lower passage for drainage, before the work at the higher level was completed, when, doubtless with a catastrophic rush, all the water still standing higher was released in a torrent until the lake level was stabilized for a time at the new level, its wind-driven waves in the meantime carving another beach in the rocky glen-slope to mark it.

19. The 'Parallel Roads' (wave-cut beaches of a former ice-dammed lake). Glen Roy, Inverness-shire.

VARVES

Lakes fed with drainage from a retreating ice-front rapidly accumulate sediments brought in by meltwater-streams. When these are flowing strongly in spring and summer they transport stones and sand and silt. The coarsest particles, rolled rather than suspended by the current, form a steep-fronted delta at the mouth and this grows outwards into deeper water. Sand, silt and clay, in suspension together are carried further out and are segregated on the bottom of deep water as layers of graded sediment, owing to their different rates of settlement. In a given place the spring and summer melt will deposit sand. Slowing current in autumn contributes only silty, finer, material while at winter freeze-up no new sediment at all may reach the lake.

Beneath the ice of winter the season's clay will slowly and gently fall to the bottom to lie upon the sand and silt already there. Renewed stream-action in the spring will bring another rush of clean, pale sand to cover the dark, even surface of the clay, so that each year's contribution of sediments, called a *varve*, stands out clearly from the preceding and succeeding year's deposits. These may be measured and counted in due course by the geochronologist seeking to describe and date the stages of glacial retreat (Fig. 20).

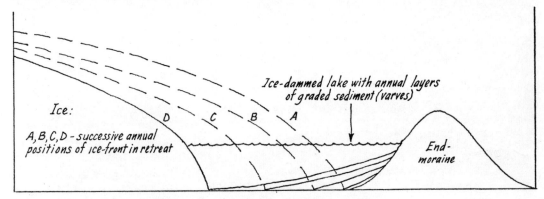

20. *Formation of varved sediments. After Zeuner (1945).*

While the retreating lowland ice-front recedes from its end-moraines and gradually lays bare the rock-basins, drumlins and eskers formed during its advance, the firn-fields in the higher mountains will shrink and the all-covering ice retreat once more into the valleys. This will leave exposed rounded-off ridges, cols and spurs between the now separate valley-glaciers. Even after complete disappearance of these, many thousands of years must pass before the signs of their prolonged action will be erased from the now bare topography they created – the semicircular high corries with tarns or lochans occupying their ice-gouged basins (Fig. 21), smoothed and striated *roches moutonnées*, blocks precariously perched on slopes and divides, truncated valley-spurs, U-shaped valleys and hanging tributaries with waterfalls. The valley streams may, indeed, have since incised some shallow V-shaped cuts into the valley-floors, between the time of the glaciation and our own day (Fig. 22). Renewed frost-weathering of peaks and precipices may have formed scree-slopes at their feet, covering minor ice-rounded features. Vegetation may have clothed the slopes where

41

21. Corrie and lake in rock-basin. Glaslyn, North Wales.

new soil, formed by chemical weathering of the rocks, has filled pockets, fissures and depressions. The changed conditions will all have been favouring the work of new, untiring geological agencies tending to obliterate traces of former glaciation. A mere ten thousand years, however, is all too little and, with us, the signs of the latest glaciation remain clearly still for all to see.

In arctic and sub-arctic climatic zones, mere abrasion and frost-cracking reduce boulders to flour-like dust without materially changing its substance. Under a more temperate climate and with adequate vegetation-cover, chemical decomposition of the rock-forming minerals comes to the fore. Among the common minerals, quartz is little changed in cool climates, though it may dissolve appreciably in the moist

22. V-shaped modern stream valley cut into an ice-smoothed landscape. Brecon Beacons, South Wales.

tropics. Felspars and ferro-magnesian minerals, which form the bulk of the remainder of common rocks, are changed by weathering to clay minerals and more or less hydrated iron oxides and the solubles formed during their breakdown are forthwith carried down to the sea in surface run-off and sub-surface drainage. The relatively insoluble residues may be locally accumulated and prevented for the time being from being washed away also. The most effective stabilizing agent for such *soils* is the root-systems, sometimes forming a dense mat of fibrous roots near the surface, of the plants which colonize them and draw from them the mineral nutrients which they need for their growth. Ultimately, however, soil, like all other finely particulate materials resulting from the processes of weathering, is liable to be washed or blown away, in a moist climate mostly by rainwash and gravity, whereby it finds its way into streams and is carried further towards its eventual resting-place – the sea.

Such sediments may not immediately reach salt water, especially across a recently-glaciated landscape. Any check to the transporting current will enable its load of rock-particles and soil to fall out of suspension. A major source of such a check in a river's course would be a lake (Figs. 23, 24), receiving, as a growing delta at its head, the coarser part of the stream's load and collecting in its deeper parts all the finer sediments which take longer to fall out of suspension. After rain in the mountains, or with spring thaw causing a rush of snow-water from the higher valleys, the

23. Stream-delta at head of Derwentwater, Cumberland.

upper Rhône brings a turbid flood into the head of Lake Geneva, or Léman, near Montreux. The overflowing lower Rhône at Geneva, however, is green and crystal-clear, all the suspended mud having sunk, across the kilometres of still water intervening, to add to the sediments which, in time, will completely fill the lake-basin. Already an almost horizontally-lying reed-covered delta extends between 6 and 10 kilometres across the headward part of what was once open water, at Villeneuve (significant name!). The Léman is still some 60 kilometres long between Villeneuve and Geneva and will last as a lake for a long time yet, but many smaller lakes are disappearing, or have already vanished. The rivers now meandering across their levelled basins, if one day 'rejuvenated' by crustal movement or the arrival from

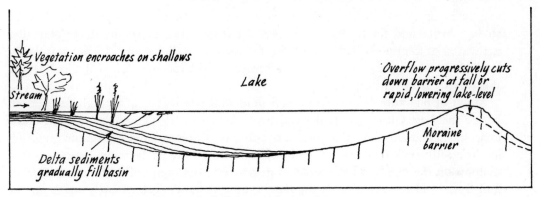

24. River-delta, lake and overflow-notch. Sediments and vegetation.

below of an active knickpoint, may begin cutting down into the accumulated sediments and starting them on a further stage of their journey to the sea.

Not only silting with rock materials brought by affluent streams is responsible for the gradual filling of lake-basins. In interglacial conditions, that is under a temperate or even warm climate, a lake will favour the development of a rich vegetation. Natural still waters support a plentiful flora and fauna, and, if the individual plants and animals are of microscopic dimensions, their astronomical numbers make their contribution to sediment-accumulation an important one. Some of the species may possess calcareous (algae) or siliceous (diatoms) skeletons which, on occasion may form beds of sediment consisting of practically nothing else – lake-marl and diatomite respectively. Among macroscopic plants, totally submerged pond-weeds, plants with floating leaves and flowers, like water-lilies, reeds and sedges rooted under water but with emergent stems and flowering heads and plant-colonists of marshy soils at lake margins – all contribute organic and mineral residues when they die to underwater soils and sediments.

At the bottom of even a shallow lake there is, because of this mass of decaying matter, some shortage of oxygen, so that many processes of aerobic bacterial decomposition reach a point at which they can advance no further. Humus thus accumulates and this, together with the remains of planktonic organisms and the mineral sediments brought by streams causes the open water to shrink from the margins, inwards. A lake, left to itself, inevitably dwindles, through the stages from open water, through weed-tangle, reed-swamp and bog to a relatively firm, at least

45

seasonally dry, land surface. This is colonized, first by grasses and herbs, then by woody shrubs and finally by forest trees. In larger, and especially deep, glacially-excavated or dammed lakes, this obviously takes a very long time, by our standards, but lakes, as features of a glacially-formed landscape are, in geological terms, short-lived, however large.

Much of our knowledge of the environmental conditions of interglacial periods before the Last Glaciation is derived from studies of the deposits of ancient lakes, long since filled and dry. The special conditions of preservation of organic materials in their sediments leave us recognizable remains, ranging from pollen-grains and diatoms on the one hand to trunks of great trees, the bones of large animals, birds and fishes and the remains of the dwellings and handiwork of early man, on the other.

RIVER-DEPOSITS

If lakes fill up and disappear under a temperate climatic régime, rivers, on the whole, excavate their valleys in the same conditions. We have seen how, in a glacial climate, periglacial streams flow for only part of the year and are so overloaded with frost-weathered rock material from their valley-slopes or parent glacier-fronts that they continuously build up (aggrade) their beds and valleys. Now once again flowing strongly during most of the year, and especially so in winter and spring, with heavy rain over their catchments and snow-meltwater, they are able to cut into their valley-floors and eventually transport all the eroded materials down to the sea. Vegetation on the valley-sides holds and stabilizes the formerly loose screes and solifluction-fans with its root-mat, chemical weathering and the leaf-fall from the plant-cover contribute to soil-formation and slope washing of mineral materials is almost halted. When there is plentiful water to supply the river, the racing current in its channel enables it to erode strongly, its transporting power for rocks and sediments increasing as the square of the current-velocity. This means that a current twice as fast as before will move and transport stones four times as massive. Thus, in time of maximum flood, even large boulders may be moved, not, perhaps bodily suspended by the water, but rolling and bounding along, crashing and grinding into and over obstacles, breaking and bruising everything in their path. So, too, for smaller cobbles and stones: temperate-climate river-action tends to break off prominences, to grind them on all sides and to turn them into smooth pebbles – in complete contrast with the scarcely rounded, but striated glacial and periglacial gravel-stones. Only in flood-time does significant erosion and movement of coarser material take place. Clay, silt and even sand may be transported by even a moderate current, but inspection of the gravel bed of a clean stream in summer will show, by the slimy growth of algae that covers their exposed surfaces, that they are not now being moved, nor have done so for some time. Nevertheless, in the long run, any river in a moist-temperate environment is, on balance, continually lowering its bed by cutting into it at intervals and carrying off more rock-material than is supplied laterally by its valley-sides and tributaries. This is the case, at least, in its upper and middle reaches.

This result follows: that after the retreat of a glaciation which caused a valley to be aggraded, the renewed erosive power of the river will make it begin to cut a deep channel into the valley-filling (Fig. 25). As the result of the widening and downstream migration of meanders, in time, nearly all of the pre-existing sediments will be cleared out from bluff to bluff, and even new cuts into bedrock be made, until the stream reaches its base-level. The new valley will, unlike one that has recently held a glacier, have a transverse V-section. In certain places, however, by the mere chances of location and timing of the passage of meanders, small areas of the former fill may escape erosion, until the river has cut its valley-bottom so far down that, even in flood, it can no longer reach these remnants. They will then be preserved as high terraces (*'climatic' terraces*), of which the composition, arrangement and fossil contents (if any) will clearly bespeak their accumulation in a preceding cold period. The summit of such a terrace, if not too small a fragment or too much denuded subsequently, will mark the level of the former floodplain. Even if denuded, it will at least yield a minimum figure for this.

SOILS

With the onset of warmer conditions and the spreading of a cover of vegetation, chemical weathering of rocks *in situ* and of glacial drifts mantling the countryside will begin.

Rain-water contains gases, dissolved from the atmosphere, which render it a better solvent and oxidant of mineral substances than pure water. Not only are

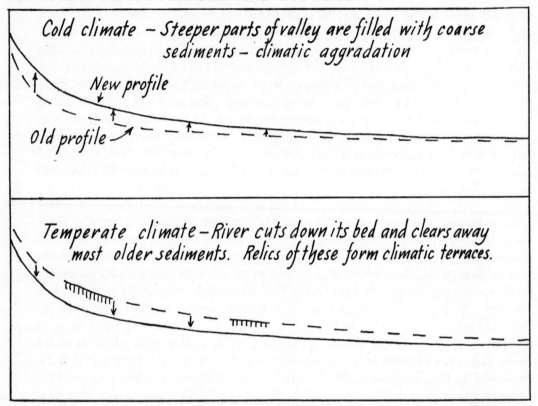

Cold climate – Steeper parts of valley are filled with coarse
sediments – climatic aggradation

New profile

Old profile

Temperate climate – River cuts down its bed and clears away
most older sediments. Relics of these form climatic terraces.

25. Formation of climatic river-terraces.

inorganic chemical changes favoured by higher temperatures and the year-round
availability of moisture, but the organic material of plants, once they have flowered
and seeded and died, will be returned to the maturing soil as humus, enriching it as
a substrate for further plant-growth. Humus further increases the acidity and dis-
solving power of percolating water, so that the colonization of a land-surface by
plants assists the further development of the soil.

Though extremely slow, these processes ultimately degrade even hard rocks into
fine-grade materials, mainly highly siliceous and stained more or less with iron oxides,
the water-solubles being carried away in the drainage. A land-surface thus subjected
to continued chemical weathering becomes greatly changed in comparison with the

fresh, unweathered parent material of the soil which now covers the surface. Should conditions once more change and a fresh layer of sediment of any kind be laid down over the weathered surface without unduly disturbing it, the soil will be preserved, relatively unchanged thereafter, as a 'fossil'.

Soils today, to be observed in active development at the land-surface, under known climatic and other environmental conditions, provide us with information which can be used to interpret the fossil soils attributable to past temperate intervals during the Pleistocene. In particular (though by no means exclusively), soils developed on periglacial loesses during periods of glacial recession and full interglacial periods, have proved very instructive in elucidating the sequence of soil-events in relation to repeated phases of loess-formation, showing something about the characteristics of the climate under which they were formed (Fig. 26).

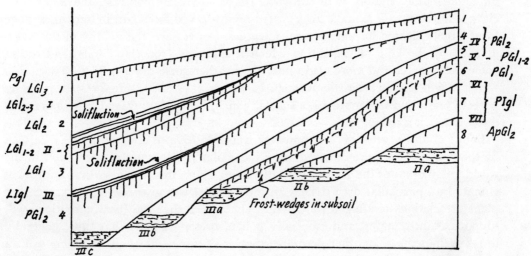

26. Sedlec, Czechoslovakia: sequence of 8 loesses and VII fossil soils on terraces of River Vltava. After Prošek & Ložek (1957).

Other things being equal (a rather large assumption, in some cases!), the depths below surface to which loesses and other sediments have been chemically altered during a succeeding period of milder climate, gives some measure, if a rough one, for comparison of the intensity and duration of the weathering-conditions to which they have been subject. Not only, therefore, are ancient soils of value for suggesting

D

49

the general character, as to climate, of their contemporary environment, but their occurrence in the stratigraphical succession of Pleistocene loesses, these corresponding to repeated periglacial conditions during the glaciations, show the number and sequence of more temperate intervals which have affected the particular area, and so are of great chronological value.

VEGETATION

The nature of the vegetation which will grow on it under a temperate climate depends to some extent on that of the soil. We all know areas of poor, sandy soils in Britain which today present a bleak moorland environment, or, at best, are suitable for afforestation, mainly with softwood timber-trees. By contrast, the loams and clays of lowland valleys make for a varied broad-leaved forest, or, in our times, have been largely cleared for agricultural purposes. Conversely, the quality of the soil is affected by the kind of vegetation which it supports – moorland soils tend to become even poorer and more acid under the heather-gorse-bracken plant association which now is so characteristic of them. In earlier days, they were forested, in many parts at least, and under trees bore a soil which, if not rich, was at least more productive than it is today.

Both soil and vegetation are, however, very sensitive to climate, as may be seen by the way in which the distribution of particular major types of vegetation in the world follows that of the Earth's climatic zones. From this it follows that improving climate in a particular area during an inter-glacial will give rise to a succession of plant associations tending towards the richest flora that the local conditions will support. Plant-remains, and especially pollen, preserved from that time, generally in lake-sediments or in other permanently wet situations, thus give us some overall view of the vegetation of the period, enabling us to deduce the character of the contemporary climate. This may often prove to have been different from that prevailing today at the place in question.

Even in well-aerated situations, such as a fossil soil, if the soil-reaction is not too alkaline, pollen may be preserved for examination with the same end in view, but, in the case of loesses, the chemical environment is almost invariably calcareous, so that pollen cannot generally be recovered. The method has been applied with success to other buried soils of a more acid nature.

ANIMALS AND MAN

Since all animals ultimately depend on green plants for their nourishment (though carnivores do so indirectly, through their prey, which are generally direct vegetable-feeders), the available plant-assemblage will largely determine the species of animals which can exploit it. Even the environment of the periglacial zone during a glaciation sufficed to support animals surprisingly numerous and varied in nature. This was the case in summer at least, for winter-long snow-cover and arctic winds make intolerable conditions for nearly all, even the most hardy animal species, many of which must either hibernate in shelter or migrate seasonally.

The richer vegetation of the temperate zone during an interglacial permitted the development or immigration of many of the less highly cold-adapted animal species and of these, in his turn, man was often the companion and predator. At least during most of the Pleistocene, man, sprung from a tropical-forest-living progenitor, was incapable of surviving in glacial or periglacial conditions. Latterly, as even more in the present, he did develop and adopt cultural devices to enable him to survive, and indeed thrive, in cold climates. Thus, for the greater part of the Ice Ages, man was exclusively an interglacial or interstadial visitor in the present-day temperate zone. During the Last Glaciation the Neanderthal group of human beings had learnt the use of fire and probably of skin clothing. They certainly sheltered in caves and so were enabled to survive, hunting the reindeer and other loess-steppe animals of the periglacial zone.

SEA-LEVELS

Melting ice-sheets everywhere restore floods of water, by streams and rivers, to the oceans. It is calculated that, were the surviving ice-sheets of Greenland and Antarctica to melt away completely today, the general level of the oceans would be raised, relative to the continents, by some 100 m. (330 ft.) this is called the *eustatic effect*.

At the height of a maximum glaciation in the Pleistocene ice covered northern Europe down to the 50th, North America in places down to the 40th parallel of latitude. Over the centre of ice-accumulation in Scandinavia it stood some thousands of metres thick. Even in the west of Scotland only peaks over 3,000 ft. in height (about 900 m.) stood above it as nunataks. So vast a volume of water abstracted from the oceans may have lowered the general sea-level by close on 600 ft. (180 m.) and this would have laid bare most of the bed of the North Sea, eliminated the English Channel, and even the Irish Sea, save for a string of lakes. So also for all other sufficiently shallow continental shelves. Instead of the nearest sea-coast to London (apart from the Thames estuary) being a mere 50 miles away, at Brighton, it would then have been about 200 miles west of Land's End (Map, Fig. 17).

Quite apart from the proximity of the ice-edge (at Hampstead, and even reaching the Thames at Hornchurch in Essex) this would have given a winter climate in south-east England equivalent to that of central Germany at the present day.

Reversing the process: re-flooding the continental shelves and even raising the sea-level somewhat beyond its present height would have the converse effect of making the climate of north-west Europe once more oceanic and equable, with winter frost a rarity rather than the rule.

The shrinking and eventual breaking up of the glacial anticyclone, when the ice-sheet had melted back to its critical size, would once more permit the present, more northerly tracks of Atlantic 'lows', bringing southerly warmer air-streams to British latitudes.

Just as a falling sea-level during an advancing glaciation 'rejuvenates' rivers by giving them steeper gradients in their lower courses and making them cut deeper channels into valley-floors, so one that is rising dams back the water at their mouths. The consequent check in current velocity causes sediments, in the process of being transported, to fall out of suspension, so that rivers, the lower reaches of which are affected in this way, gradually aggrade their courses and estuaries and build their flood-plains up to the height of the new sea-level (*eustatic aggradation*) (Fig. 18).

Since the climate during such an aggradation is improving, perhaps becoming even warmer than that of today, contemporary plant and animal remains preserved in the growing deposits will be of a temperate-climate character, in contrast to those found in climatic terraces occurring higher up the course of the very same river.

There will be a region, in the middle reaches of a river, where, depending on the precise heights of successive interglacial sea-levels, and those of knickpoints started during the intervening periods of low glacial seas, in which climatic and eustatic aggradations will perhaps interleave in a rather complicated manner. In the part of a river's lower course, within the control of the changing sea-level, aggradation-deposits will, on the whole correspond with warm stages, of renewed down-cutting with cold climatic conditions. Since the sea-level during the last 0·3 million years seems never to have stood at more than about 100 ft. (33 m.) above its present height, in the case of large rivers, like the Rhine or the Danube, no part of the river's course at present above that level can have been eustatically affected during the period in question – at least as regards aggradation.

Any knickpoints which started cutting as a result of the low glacial seas, and which had already so far progressed upstream as to escape being overtaken by subsequent high interglacial sea-levels, will continue their work, registering downstream of their present positions the rejuvenated long-profiles of their times.

LAND-BRIDGES

While low sea-levels, caused by extensive continental glaciations, may open up land-bridges between continents and reunite offshore islands with their mainlands, they seem never to have exceeded the 200 m. (600 ft.) submarine contour. Thus, though on several occasions Britain has ceased to be an island, north-eastern Asia was united by a wide land-bridge to Alaska and much of the South China Sea and western Indonesia formed part of Asia, certain important seaways which exist today were never bridged in the Pleistocene. Among them are those in the western and central Mediterranean (Gibraltar-Morocco and Sicily-Tunisia), those between the West Indian islands (save Trinidad) and North and South America respectively. Nor were those between Borneo and the Celebes and between the islands of Bali and Lombok in the Sumatra-Java island chain. New Guinea, on the other hand, would have been joined to Australia and the Japanese islands to Korea; Ceylon to India and southern Arabia to north-east Africa. Madagascar, separated from Africa by the deep Mozambique Channel, has probably not had any land-connection with that Continent since the earlier part of the Tertiary Era.

ISOSTATIC RECOVERY

The growth of a large sheet of land-ice loads the crust of the earth locally, and, in time, depresses it (isostatic effect) in relation to the general level of the surrounding ice-free areas. Where the load-stress is greatest, below the thickest part of the ice-

53

sheet at its centre of accumulation, the sinking of the crust is greatest, so that the former land-surface becomes 'dished' beneath the ice and may present, as in Greenland today, a great basin with its centre actually below sea-level, while its only lightly-loaded margins still stand high above it (Fig. 27).

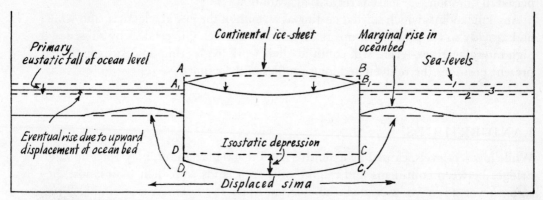

27. Sialic continental block, floating in sima, deformed and depressed by ice-loading. After Fairbridge (1961).

On waning of the ice a slow recovery, 'recoil', begins and may continue long after its final disappearance, the time-lag of thousands of years being due to the very viscous flow and readjustment of mantle-material below the crust, which had been displaced, equally slowly, during the accumulation of the ice-load now removed. The central area of the Last Glaciation ice-sheet in north-east Sweden and the north of the Gulf of Bothnia is, today, rising from this delayed recoil by nearly 1 cm. per year, though practically the whole load of land-ice was removed about ten thousand years ago.

Such slow and uneven recovery of the level of the land-surface after glaciation evidently affects the relative positions of land and sea in the immediate glaciated areas. The effects may be seen, for instance, in the west of Scotland, where, since the maximum of the Last Glaciation, the land has, in places, risen by some 100 ft. (33 m.) and marine platforms on the coast, cut by wave-action in early Postglacial times, now stand dry at about this height above present-day sea-level. These isostatically raised beaches are correctly so called, in contrast with high beaches which mark ancient interglacial sea-levels far from glaciated areas which are eustatic in origin. These now stand above present sea-level, not because the land has risen, but because the sea-level has fallen since they were formed.

Obviously, in areas marginal to former ice-sheets both influences have, at different times, been at work, so that interpretation and dating here of beaches and river-terraces of earlier times by their heights is complicated unless the net effect of the two contrary movements of land and sea can be estimated, or the influence of one or the other safely be discounted.

The margin of important isostatic uplift since the Last Glaciation (zero isobase) runs through the Great Lakes area in North America, approximately bisecting Lakes Michigan and Erie and tending on a somewhat more northerly bearing both to the west and to the east. In north-western Europe it lies approximately at the England-Scotland Border and passes through north Jutland and the south of Sweden.

Where modern coastlines and the raised beaches run across the isobases, therefore, the raised beaches will be more or less tilted, owing to differences in the amount of isostatic recovery experienced by their different parts since they were cut.

Apart from isostatic sea-level fluctuations, incidence of a glaciation in the north temperate zone has other worldwide effects on climate. We have seen that the establishment of a stable anticyclone over the glaciated area diverts far to the south the oceanic depressions which at present reach north-west Europe from the south and west. At a glacial maximum the successive 'lows' in the western approaches would experience a collision with the air-currents of predominantly north-eastern trend radiating from Scandinavia and the Baltic, and be forced southwards. A depression which, today, might cross the British Isles en route for Norway would be turned aside, across the Bay of Biscay, or even along the shores of north Africa, shedding its moisture not on the Western Isles, but on the Canaries, the Atlas Mountains and the Saharan ranges, rather than over northern and central Europe. Evidently, by this change, with the subarctic zone extending to the Pyrenees and the Apennines, the temperate climatic zone of all-the-year-round moisture would invade the desert zone at present centred on the Tropic of Cancer and give much of now arid north Africa a climate of at least Mediterranean character.

In the Equatorial Zone one can conceive of no fundamental change at any time in the perennial heat and moisture, so that the spreading of Mediterranean conditions down to perhaps Lat. 30°N would seem to have taken place mainly at the expense of the desert and perhaps the more southerly savannah belt.

E. A. Bernard (1962), however, has pointed out that the cooling of the oceans by a northern circumpolar glaciation extending far into the present temperate zone would not be confined to extra-tropical areas. Ocean currents would inevitably convey considerable additional cold water from more northerly sources into the tropics, and even into the Equatorial Zone. Since rainfall here is orographic (dependent on the presence of high ground) and convectional (caused by updraughts of air-columns over land areas strongly heated daily by an almost vertical sun) and since the humidity of that atmosphere is ultimately gained by evaporation from the surface of an adjacent warm ocean, any important fall in temperature of that ocean would reduce evaporation and with it the humidity of air-masses and the rainfall reaching equatorial coasts. He is of the opinion, therefore, that in sub-Saharan Africa, at least, wet periods (pluvials) correspond with interglacials in the north and drier periods with the glaciations – not the other way about, as had generally been thought hitherto. The principle will apply also in the humid tropics in other continents.

CHAPTER 3

Outline history of the Pleistocene

Part I: Lower and Middle Pleistocene

The Pleistocene Period is now reckoned to be some 2–3 million years long, including a long earlier portion known as the Villafranchian, which was formerly assigned to the Pliocene but, by one definition, that of Haug (1911), to be included in the Pleistocene because its deposits contain fossil remains of true horses, cattle, elephants and camels. The assimilation of the Villafranchian to the Pleistocene was only officially approved by the International Geological Congress meeting in London in 1948.

In addition to the newcomers specified by Haug, Villafranchian deposits contain members of many archaic mammalian genera, survivors of the Pliocene, and, certainly at times a warm-temperate, indeed sub-tropical, forest flora, very different from that of the present day. Despite periodically greatly lowered ambient temperatures, therefore, the Villafranchian represents, on the whole, a continuation of the later Tertiary conditions, unmarked by widespread sudden breaks.

The later Cretaceous had been a time of almost world-wide warm, equable climate and continental submergence, as is clear from the large areas of the continents of today which are covered by the Chalk and other Cretaceous formations. Throughout the Tertiary, marine temperatures (Emiliani, 1958) had been steadily falling, so that, by the end of the Pliocene, the northern oceans and seas, though still warmer than they are today, had become much cooler than they had been for the previous 70 million years at least.

In part, certainly, this was the result of increasing continentality, due to widespread land-uplift in the Oligocene and Miocene. These were the Periods during which the latest major phases of folding and mountain building in the Earth's history (Alpine orogeny) took place. The Alps, the Himalayas and the American Cordilleras, among other geologically 'young' mountain ranges and massifs were raised in these phases, Cretaceous fossiliferous marine rocks being found near the summit of Mount Everest, for instance. The breaking of communications between

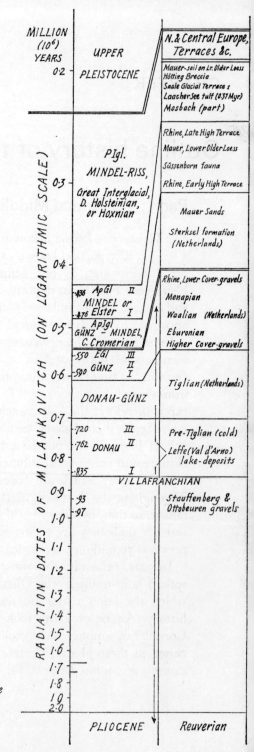

28. *Correlation-table for Lower Pleistocene and Middle Pleistocene, showing palaeomagnetic epochs.*

Sea-levels	Palaeomagnetism	MIDDLE Thames, S.E.England	PLEISTOCENE E.Anglia, Midlands	N.America	Hominids
Late, 32m. Tyrrhenian Early, 45m.	EPOCH	BoynHill, Handboro Terrs. Slindon-Goodwood beach Kingston Leaf & Dartford Heath gravels	Hoxne lake-deposits Swanscombe, Clacton gravels Nechells (B'ham) Kirmington (Lincs)	Yarmouth Interglacial	Swanscombe & Steinheim _H.sapiens s/spp._
? −100m.		Black Park Terrace	Hornchurch, Maldon, Lowestoft boulder clays Cromer Till, Norwich brick- earth, Chiltern (western) Drift	Kansan	_H. erectus s/spp:_ Choukoutien, Ternifine, Vértesszöllös Mauer
Milazzian (60m.)	NORMAL	Ambersham Terrace Winter Hill Terrace Harefield Terrace	Cromer Forest Bed	Aftonian Interglacial	_H. erectus erectus,_ Trinil, Java
(low)	BRUNHES	LOWER Harefield bench cut Higher Gravel-Train aggradation H.G.T. bench cut	PLEISTOCENE Beestonian (cold) Pastonian (temperate) Baventian (cool)	Nebraskan	
80 m.			? Antian (temperate)		
100 m.		Weybourne Crag			
Sicilian 105 m.	EPOCH	?	Thurnian (cool)		
Jaramillo Normal Event	0·90 1·0 REVERSED	?		0·87 Bishop Tuff, on (K/A) ? pre-Nebraskan boulder-clay	_H. erectus leakeyi_ (Olduvai Bed II)
(Unnamed) 129m.	MATUYAMA REVERSED	Hertfordshire Pebble- Gravel Norwich Crag ? Westleton Beds ?	Ludhamian (temperate)		
Calabrian 180m.		Newer Red Crag Lenham Beds			
Gilsa Normal Event					1·75 _H. habilis_ at (K/A) Olduvai, Bed I
		Older Red Crag			
		Coralline Crag			

59

tropical and more northern seas by the rising land-masses interrupted free heat-exchange between latitudinal zones by marine currents and so introduced greater temperature-differences between them.

By perhaps three million years ago, the sub-tropical sea which had laid down the Coralline Crag in East Anglia gave way to that of the Red Crag, a ferruginous, sandy shallow-water deposit, and, in this, many southern species of molluscs began to dwindle and become extinct, while temperate species still inhabiting British waters slowly increased and the immigration of even some northern, cold-water, forms began. Zeuner (1945, p. 106) showed that two distinct influxes of cold species took place, in the times of the Newer Red Crag and Weybourne Crag stages respectively. He interpreted these as indicating the occurrence of at least two cold climatic phases before the time of the Cromer Interglacial (see below). The former of them (Red Crag) seems to be contemporary with a 200-m. (660 ft.) high sea-level (Fig. 39), called – from its Mediterranean occurrence – the Calabrian, which, there too, shows the first influx of a cold marine fauna. The shoreline gravels of the Diestian Sea, which extended at this time from the Netherlands into the London Basin, have been mapped at levels close to the 600-ft. contour near the summits of the North Downs and those of the Chilterns (Wooldridge, 1927). These would have a date of about 1·5 m.y.a. The Weybourne Crag might be of only half that age. The intervening Norwich Crag contains drifted remains of a forest and parkland mammalian fauna (Fig. 28).

In the Netherlands, similarly, the Coralline Sea gave place to a cooler marine stage, corresponding to the Red Crag, and this was followed by the Tegelen Series of sands and clays, again with a forest flora and fauna. Twice renewed cold is evidenced in the Kedichem Series, overlying the Tiglian, which is interrupted by a milder phase, called the Waalian Interglacial (Fig. 28).

A long succession of lacustrine Villafranchian deposits is found at Leffe, and elsewhere in the Val d'Arno in Tuscany, and these provide both plant-pollen and mammalian remains. They consist of repeated thick beds of laminated calcareous clays, alternating with thinner peaty or lignite (woody) layers. Their pollen contents show that the former correspond to colder, the latter to warmer climatic oscillations. As in Holland, three major cold periods stand out and these may represent the effects of the same three glacial events. There are, in addition, two lesser cool oscillations suggested by the pollen-spectra, and these are harder to correlate with the less detailed North Sea regional succession (Fig. 29).

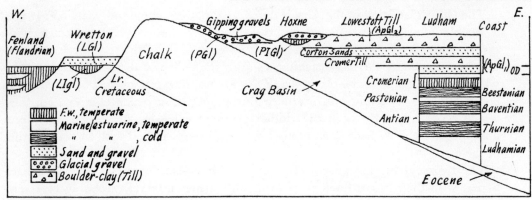

29. Pleistocene of East Anglia: diagrammatic W–E section (length 60 miles). After West (1968).

Most of this sequence long precedes the first direct evidence of glaciation in northern Europe, though it undoubtedly points to recurring intrusion of colder conditions, with, however, equally distinct long warm intervals. The Tegelen clays, for instance, exhibit remains of species of trees like *Magnolia*, *Liriodendron* (tulip-tree) and *Cinnamomum* (cinnamon) which survive naturally today only in the much warmer climates of south-east Asia and the southern United States – Pliocene survivals which have never since the Ice Age recolonized Europe, though they can survive here perfectly well when imported and grown in gardens, though they may not set viable seed.

In the Alps, the first evidences of local ice-caps spreading out across the north-Alpine foreland into the Danube basin, were distinguished by Eberl (1928, 1930). At least three distinct ice-advances are represented, known as the Donau (Danube) Glaciations. They perhaps correspond with the cold marine molluscan faunas of Mediterranean high beaches, approximately 100-m. (330 ft.), called the Sicilian. The Tegelen Series, from near Limburg in the Netherlands, probably falls into the temperate interval, called after it 'Tiglian', between the Donau group of Alpine ice-advances and the classical 'First' glaciation of the Alps, Günz, the earliest of the four recognized by Penck & Brückner (1909). Palaeotemperatures calculated from ocean-bed deposits recovered in deep-sea cores in the tropical Atlantic and elsewhere (Ericson & Wollin, 1966) confirm the occurrence of a number of Early Pleistocene cool phases, of which the dates proposed remain controversial, but which may correspond to the Donau events of Eberl.

The Günz glaciation of the Alps is dated to about 600 thousand years B.P. It was a double advance, represented by two moraines, respectively connected with distinct spreads of glacial outwash-gravels lying at slightly different levels. Neither advance-phase was very far-reaching, even in the valley of the north-Alpine river which gave its name to the glaciation. They are not now represented by any known continental glacial features in northern Europe, for, if any such ever existed, they would repeatedly have been overridden by more extensive subsequent ice-sheets and almost certainly have been ploughed up and redistributed with the other drifts of these.

The two cold marine phases (Eburonian, Menapian) of the Dutch Kedichem Series, with their intercalated brief Waalian warmer interval, probably represent the periglacial effects in the North Sea of two Scandinavian equivalents of Günz I and II and their Interstadial. Though these fragmentary and scattered traces show that for quite some time (perhaps 50 thousand years) northern and central Europe became distinctly cold, with an Alpine ice-cap for certain and another probably centred over northern Norway and Sweden, Günz does not seem to have had much influence further afield, for the succeeding Antepenultimate (Günz-Mindel or Cromerian) full Interglacial shows the persistence of at least a few mammalian survivals from the Pliocene, alongside some typically Pleistocene newcomers.

The North American correlative of European Günz was generally supposed to be the Nebraskan glaciation. According to Flint (1947), this extended south and west to Nebraska, Kansas, Iowa and Missouri and probably was little less extensive than subsequent glacial drift-sheets, though practically only in this one area exposed by erosion in river-valleys. This looks rather unlike Günz which was neither comparably extensive with the rest of the glaciations nor has survived at all in the corresponding situation in northern Europe. A recently-obtained Potassium/Argon date of 870 thousand years B.P. for a volcanic explosion-ash, the Bishop Tuff, which directly overlies a boulder-clay provisionally identified as Nebraskan, throws considerable doubt on the Günz-Nebraskan correlation. Either the name Nebraskan is here mistakenly applied to a hitherto-unrecognized ancient boulder-clay, older by some 270 thousand years than European Günz, or, if it is correct, there is no possible correlation between Nebraskan and Günz. The date given would, in fact, more readily be applicable to the Donau glaciations of Eberl, though it is a little early even for this, according to current European chronological ideas. It is most unlikely that the Potassium/Argon date could be as far out as this, for the method, as we shall see

(p. 145), has given most acceptable dates for even later volcanic events in Europe.

The succeeding Interglacial, Günz-Mindel of the Alps (Cromerian in north-west Europe, Aftonian in North America) is represented by several famous fossiliferous sites in Europe. The Cromer Forest-bed Series, typically displayed on the East Anglian coast, in cliff-sections, gives its name to the interval, which was about 75 thousand years long (Fig. 29).

The Series is composed of a number of sandy, sometimes peaty, fluviatile to estuarine deposits laid down in the course of a large river flowing in a generally northerly direction, presumably into the North Sea. Though the beds are now at about sea-level, it must be remembered that the North Sea basin is a geosyncline, continually sinking slowly as it fills up with sediments, so that the Cromer river was surely flowing at a much higher level than this when its deposits were first laid down. The hinge-line of the geosynclinal down-warping runs about north-south through Braintree, in Essex, so that everything to the east of this has been more or less depressed in relation to the area to the west of it, which has remained unaffected. An interglacial sea-level at about 200 ft. (55–60 m.), called the Milazzian Sea in the Mediterranean area, is represented by many erosion-surfaces in south-eastern England (Wooldridge, 1928) corresponding with the Winter Hill Terrace of the middle Thames, so that the Cromer Forest Bed, in its present situation, has clearly moved out of its original altitudinal position.

The Cromer Forest Bed has yielded a rich fauna and flora and pollen-analysis has given spectra characteristic of this Interglacial in Britain. Typical sites of comparable age on the Continent are the 45-m. terrace of the Somme, at Abbeville; Süssenborn, near Weimar in Thuringia; Mosbach, near Mainz and Mauer, near Heidelberg, both in the Rhine valley, all of which are in river-sediments.

The faunas display a family likeness, with a few Villafranchian survivors still, but an assemblage of animals (though some now extinct) in general with an essentially modern appearance. The flora, too, is temperate, including none of the sub-tropical Villafranchian species, although those typical of the Interglacial pollen-spectra in Britain now survive only in eastern Asia and the Americas or the more eastern parts of the continent of Europe and have not been represented here since the Last Glaciation. It is clear, therefore, that the Günz cold periods never made a clean sweep of the pre-existing plants and animals, but that, even if they had been driven southwards by the advancing cold, some of them at least had found suitable refuge-areas from which to re-colonize the more northern parts of the Continent when the ice retreated. Several species were, nevertheless, making their final appearances in the Cromerian Interglacial, so that their remains, when recognized, may be used with confidence to assign a deposit of unknown age to this, or to some earlier, period.

Man, in the form of the long-famous Mauer jaw, appears now for the first time in Europe. This isolated fossil would consort suitably with a skull of the *Homo erectus* type (ex *Pithecanthropus*) from the apparently contemporary Trinil Beds in Java, where Hominid predecessors are known, of the *Australopithecus* and *Homo* stages of development, both from the underlying Djetis Beds, which are probably of Villa-franchian age.

The Cromer Forest Bed is succeeded by a series including the Arctic Fresh-water and the marine *Leda myalis* Beds, the flora and fauna of which clearly point to a renewal of cold conditions. This marks the beginning of the Middle Pleistocene, according to Woldstedt (1958) and most subsequent authors.

The Antepenultimate Glaciation (Mindel of the Alps, Elster in northern Europe, Lowestoft in East Anglia, Kansan in North America) was a very different thing from Günz. Like the latter it was a dual advance with a comparatively slight interstadial separating the two maxima lasting under 50 thousand years. In the extent of its ice-sheets, however, it was, notably in Central Europe, the greatest glaciation of all (Fig. 30). Even in Britain one of its advances reached southwards to the Thames at Hornchurch, in Essex, and to the Bristol Channel, covering Ireland completely save for the extreme south-western tip of Co. Kerry (See map Fig. 30). It is probable that the Chalk ridge between Dover and Calais was breached at this time (if not before), for with an ice-barrier in the North Sea down to the Thames Estuary and the Netherlands, the waters of the Elbe, Weser, Rhine and Thames must have drained through the Straits, though the gap was probably nowhere nearly as wide as it is today. On the Continent, the ice-edge crossed the Netherlands, lay against the northern flanks of the Harz and Riesengebirge and, further east its ground-moraine covered a wide belt of country beyond that of the other giant glaciation, Saale. Eastwards again, it lined the southern boundary of the lowland which is today the Pripet Marshes, formed two huge south-projecting lobes in the basins of the Dnieper and Don rivers, well beyond the 50th parallel of latitude, and abutted against the northern part of the Urals.

At the fullest extent of the Alpine and Scandinavian ice-sheets (see Fig. 17), with local glaciers also in the Pyrenees and Caucasus, Europe, between 50° and 40°N. latitude was all in the periglacial zone. Even the more southward-extending peninsulas

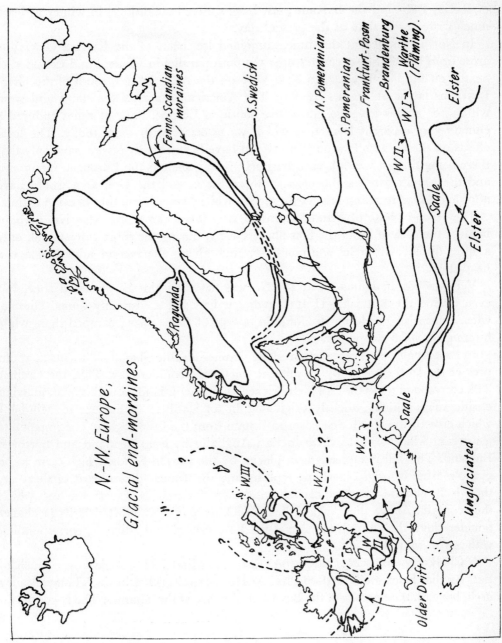

N.-W. Europe,
Glacial end-moraines

Fenno-Scandian moraines

S. Swedish

N. Pomeranian

S. Pomeranian

Frankfurt-Posen

Brandenburg

W II

W I → Warthe (Fläming).

Saale

Elster

Elster

Ragunda

Saale

W II ?

W I

W III

W II

W I

Saale

W II

Older Drift

Unglaciated

E

30. Map of glacial end-moraines in North West Europe and Britain.

of Iberia, Italy and the Balkans must have suffered conditions, in winter at least, much more severe than at the present day.

In North America at this time, conjoined ice-sheets of the Kansan Glaciation arose from three distinct centres near the 60th parallel in Canada and extended to Seattle on the Pacific coast, to New York on the Atlantic and, south of the Great Lakes, as far as St Louis (Fig. 31). The American States of the south and west, within 10° of this ice-edge, must, like southern Europe, have at times endured a climate very different from that which we associate with them today. The local effects on the flora and fauna there would have been equally severe, save only that there being no east-west physical barriers present, such as the European mountains and the Mediterranean, the most southern States and the Gulf area would have afforded refuge to many less hardy species which, here, inevitably became extinct.

The second phase of Mindel seems to have been the more extensive, perhaps because the area of ice of the first phase shrank, during the short Interstadial, only a little distance from its end-moraines, thus giving the second advance-stage a head-start.

Vértesszöllös, in Hungary, at 47½°N., was never within the glaciated zone, but even during the Mindel I–II Interstadial it had still a distinctly boreal climate, when Europe's second known Hominid, a type of *Homo erectus*, camped there while hunting the local steppe-horses.

In East Anglia, the coast sections show Cromerian Interglacial sediments at about present sea-level covered by an ancient weathered boulder-clay. This, the Cromer Till, corresponding to the Norwich Brickearth further inland, contains Scandinavian erratic rocks which evidently originated in an ice-sheet equivalent to Mindel I which crossed the North Sea basin to Britain from the Oslo region. The immediate ice-stream which brought them to East Anglia came from Scotland and northern England. The Till is succeeded, in places, by the Corton Sands, which seem to be glacio-marine in origin, perhaps representing the Interstadial retreat of sheet-ice, though the climate remained glacial as is shown by the shells of arctic molluscs which they contain. Above these, again, lies a thick deposit of Contorted Drift (a chalky boulder-clay), the ground-moraine of the Lowestoft glacial advance, corresponding with continental Elster.

A Chalky Boulder-clay (Lowestoft, Elster or Mindel II) extends over much of East Anglia, Essex and Hertfordshire. At Hornchurch, it lies in the Thames Valley itself, beneath the deposits of the Boyn Hill Terrace of the Thames. The lower of two

Greenland
Centre
Keewatin
Centre
Cordilleran
Centre
Labrador
Centre
NORTH AMERICA
Extent of maximum glaciation
Directions of rock-striae
Miles
0
500
1000
Kilometres
0
500
1000
1500
180°W
150
120
90
60
30
60
50
40
30
20
10
120
110
100
90
80°W

31. Map of glacial end-moraines in North America.

boulder-clays at Ware is probably its representative in Hertfordshire, lying at the bottom of the valley at about the 200-ft. contour.

At Mardley Heath, near Welwyn, however, and at several other places along a west-east line into Essex (Epping Forest), a still older, deeply weathered and de-calcified boulder-clay, probably originally chalky, contains *western* erratics and caps the hills at about the 400-ft. level (Wooldridge & Cornwall, 1964). This seems likely to be a correlative of the Cromer Till and Norwich Brickearth, though belonging to a different, if confluent, ice-sheet with its sources in Wales and the Midlands, which reached Hertfordshire after topping the Chiltern Hills from the north-west. It evidently corresponds also with the Hanningfield Till of Clayton (1957) which similarly caps summits between 300 and 400 ft. o.d. south east of Harlow and north of Billericay, Essex. At this time, the greater part of the Thames drainage, including that of the Wey and the Mole joining it from the western Weald, was flowing north-eastwards up the line of the present Vale of St Albans and reached the sea perhaps at the Blackwater estuary, by Maldon, in Essex. The Vale then lay at a much higher level than the present Colne valley floor, somewhere between 300 and 400 ft. o.d., while the present river crosses the 200-ft. contour some 4 miles north-east of Watford. This first (equivalent to Mindel I) ice-front crossed the Vale and gradually diverted the north-easterly drainage through low gaps south of Watford into the Finchley depression, and thence, via Ware, eastwards, as before. Later yet, even this outlet was blocked by the ice and the river was forced to take approximately its present course, south of Uxbridge and Ealing and thence to the sea via London and Gravesend.

The pre-existing valley east of Richmond doubtless contained a small stream, which may be called the 'London River', and, at the diversion, this received the whole drainage of the Thames catchment, including now the meltwater from the great ice-front to the north of London. During the Mindel I-II Interstadial, some weathering of this Western Drift boulder-clay and down-cutting of valleys took place, so that the south-west-advancing Lowestoft ice of Mindel II occupied situations at a considerably lower level. The Western or Chiltern Drift boulder-clay now survives only on the 400-ft. summits never overflowed by the ice of the second phase, which doubtless re-worked all of its other deposits which may have survived the denudation of the Interstadial. The twice-diverted Thames never regained its former northerly course, since the Vale of St Albans and the Finchley depression were, by now, both choked to too high a level by boulder-clay and outwash gravels of the two advances. During the remainder of the glaciation the lower course of the river could only begin to cut down the shallow valley of the London River, grading it to the contemporary low sea-level. This down-cutting formed the bench, at about 75 ft. O.D. at Swanscombe (see below), on which the deposits of the following Interglacial were laid down.

At the maximum glaciation, according to Flint (1947), some 32 per cent. of the world's land-area was ice-covered and the geological effects of glaciation spread, in the form of outwash-sediments and loess, hundreds of miles into unglaciated regions. The wider effect on climate was to push the warm-temperate rain-belt deep into zones now sub-arid or desertic, affecting another 10 per cent. of the land. The glacial-eustatic effect of this vast accumulation of land-ice was to lower the general sea-level by as much as 200 m. below that of the present day, exposing large areas of the continental shelves and thereby adding, temporarily, to the world's land-areas by between 2 per cent. and 3 per cent.

The 'Great Interglacial' of many European workers (Mindel-Riss in the Alps, Elster-Saale or Holsteinian in northern Europe, Hoxnian in East Anglia, Yarmouth Interglacial in North America) was a very long one. The field-evidence consists of widespread fluviatile and other terrestrial sediments and very profound weathering of glacial and periglacial deposits of the Mindel stage. Large geomorphological changes, such as deep cutting of shorelines and beach-platforms and erosion of valley-fills in inland rivers also took place. The whole aspect of these phenomena is of a very long-continued interval of intense denudation and chemical alteration of pre-existing formations under a temperate climate.

From this kind of evidence, Penck & Brückner, in 1909, estimated the duration of the Great Interglacial as being about twelve times the length of the time since the retreat of the Last Glaciation. Their estimate was a surprisingly good one when we compare it with the results of modern dating methods. It is hard for us to visualize that, on this scale, the Great Interglacial lasted nearly as long as the whole of the time since elapsed!

Within so long a period, there is plenty of room for climatic variation; and it is evident that conditions were not static all the time and, while some parts of the Great Interglacial may have been distinctly milder in middle latitudes than are the present-day conditions, not all of it, by any means, must have been so. The evidence that we have from flora and fauna embodied in deposits of this long period, suggests temperate to warm-temperate deciduous forest covering most of central and north-western Europe. It is nevertheless clear that the deposits we have, of ancient lake-

and river-sediments, each represents the history of only a relatively short, disjunct episode within that long stretch of time, and may well not be representative of the whole. The typical faunas include none of the Villafranchian survivors present in the preceding Interglacial, Günz-Mindel.

A typical and well-studied site of Great Interglacial age is that at Swanscombe, Kent, on the southern flank of the valley of the Lower Thames. It is situated at a height of about 100 ft. (33 m.) above the present flood-plain of the river, which, being tidal today below Teddington, is at about high-tide level.

A thick aggradation-series of river-gravels, sands and loams lies on a bench cut into the 'solid' (Thanet Sand, Lower Eocene) at about 75 ft. o.d. This was evidently the limit to which the river had cut its valley-bottom during the preceding glaciation. (Fig. 32).

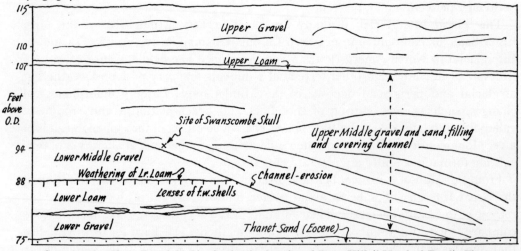

32. *Swanscombe, Kent, Barnfield Pit. Section in deposits of Boyn Hill ('Hundred-Foot') Terrace.*

The first deposit to be laid down on the bench was the Lower Gravel, about 7 ft. in thickness. Its lower part is ill sorted and loamy, with large, scarcely rolled cobbles of flint and some erratic rocks, and locally is cemented by secondary iron-pan formation. This is, apparently, a slope-solifluction deposit, only slightly re-worked and re-arranged by water. The upper part, on the other hand, is a well-stratified sandy gravel containing plentiful bones of large, temperate-climate animals, occasional humanly-worked flakes and flint cores of Clactonian type.

Over this lies perhaps 5 ft. of Lower Loam, clearly the upper floodplain deposit of a fairly mature, meandering river. Near its base it contains lenses, and fairly extensive if discontinuous, beds containing quantities of fresh-water molluscan shells, while the uppermost portion, down to 1 ft. from its summit, is somewhat redder in colour, in contrast with the yellow-brown of most of the loam. This is interpreted as a subaerial weathering-soil, denoting cessation of deposition followed by chemical alteration proceeding to some depth from the exposed surface. It is a rather immature soil, though it is completely decalcified, in comparison with the parent floodloam which is strongly calcareous, and this, alone, suggests rather intense conditions of warm-temperate moist-climate weathering imposed for a relatively short period.

Upon the Lower Loam lies another well-stratified sandy river-gravel, about 9 ft. thick, the Lower Middle Gravel, containing more bones of large mammals belonging to a forest environment, associated with flint hand-axes and flakes. This is cut through locally by a deep channel, which involves also the Lower Loam and Lower Gravel locally, and this is itself filled with cross-bedded sandy gravel – the Upper Middle Gravel. It was in this channel-filling that the famous Swanscombe Skull was found.

Following it there are another 10 ft. of more sandy sediments, apparently continuous with the channel-aggradation, and an Upper Loam, which forms the summit of the fluviatile deposits.

Over all are poorly-stratified loamy gravels and sands, over 10 ft. thick in places, being a solifluxed, slope-washed and often contorted deposit, due to a succeeding phase of cold conditions. We thus have: at the base, traces of a cold period, then three distinct phases of eustatic river aggradation under a warm climate – the Lower, Lower Middle and Upper Middle – separated by a pause in sedimentation (the soil on the Lower Loam) in the one case and a brief resumption of down-cutting between the Lower and Upper Middle stages, all covered by evidences of a renewed cold climate with frost-action.

The break represented by the soil was either quite short (if what we have is the entire soil-profile) or, possibly, very long and not susceptible of measurement (if the soil represents, as it perhaps may, only the denuded 'stump' of an original profile which was much deeper).

The very top of the fluviatile deposits lies at 107 ft. o.d. (35·4 m.), near the height of the Tyrrhenian 33-m. beach in the Mediterranean and elsewhere. This fairly confirms the date of the Swanscombe deposits and of other occurrences of the Boyn Hill (or Hundred-foot) Terrace of the Thames, to which they belong.

71

On the face of it, it seems unlikely that the entire 200 thousand years of the Inter-glacial are anywhere represented by continuous deposits. We have seen the evidence here of two breaks in sedimentation. The Lower Gravel probably falls early in the Interglacial; the Upper Middle Gravel and Upper Loam late in it. The soil perhaps records a very long break in between, during which there was no deposition of sediments at that place. It is thought that the Clacton Channel gravels, formerly to be seen at Lion Point, near Clacton, in Essex, may have been laid down during the hiatus (King & Oakley, 1935). These are fossiliferous and contain forms of imple-ments which would lie, typologically, between those of the Lower and Middle Gravels of Swanscombe. It must be remembered, however, that like the Red Crag in Suffolk, now at or near sea-level, the Clacton gravels may have been greatly depressed since their deposition by the continued sinking of the North Sea syncline, so that their true altimetric, and hence chronological relationship to the Swans-combe gravels, cannot now be accurately ascertained. There is no doubt that the Clacton Gravels belong to some part of the Great Interglacial, if not here.

Other such deposits in the Thames valley, not securely placeable in the sequence of Great Interglacial events, occur at Ilford, at Stoke Newington, at Grays in Essex and, in London itself, the Endsleigh Gardens Terrace at 80 ft. O.D. This last is probably an upstream correlative of the Stoke Newington gravels.

In the Middle Thames region, gravels of Boyn Hill, near Maidenhead and other deposits of the terrace named after this site probably also belong to this Interglacial.

At Hoxne (West, 1968) (Fig. 33) and other places in East Anglia, and the Mid-lands, as at Nechells, near Birmingham, there are ancient lake-deposits attributed to the Great Interglacial, overlying Lowestoft chalky boulder-clay, and showing a cold-warm-cold climatic sequence, as demonstrated by the pollen-spectra. The lake-sediments of Hoxne are succeeded by periglacial deposits of the Gipping (=Riss) Glaciation. The uppermost part of the interglacial deposits seems to be eroded and, here too, there is probably a very considerable disconformity – a long time-gap un-represented by sediments. On the face of it, even 6 m. depth of lake-muds at Hoxne took only a very small fraction of 200 thousand years for their deposition, so that the pollen-sequence is most unlikely to represent more than a fairly short stretch in the history of the Interglacial. Mid-Acheulean hand-axes, entirely comparable with those from the Middle Gravels of Swanscombe, have been found at Hoxne, in the upper-most lake-muds.

On the dip-slope of the South Downs at Slindon, Sussex, there is a marine beach-

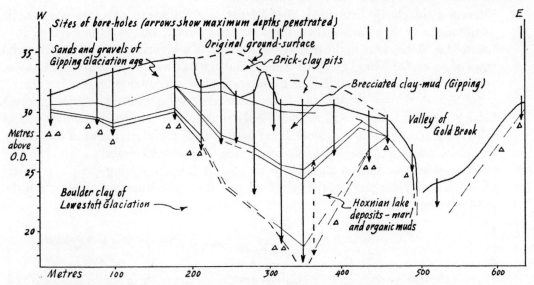

33. Hoxne, Suffolk. Profile shown by borings across site on ancient lake (Vertical scale × 15). Redrawn after West (1956).

gravel with its summit at about 100 ft. This evidently belongs to the same high sea-level (Tyrrhenian) as that which caused the Middle Gravel aggradation at Swanscombe. A solitary Middle Acheulian hand-axe comes from the surface of a sand underlying the gravel.

Across the Channel, the '30-m.' Terrace of the Somme, in northern France, shows a fauna and human implements very similar to those from the Thames valley.

At Torre in Pietra, near Rome (A. C. Blanc, 1954) – the site must date from the very beginning of the Great Interglacial, if the radiometric date by potassium/argon of 433 thousand years B.P. is correct. The presence of Chellian-type hand-axes there supports such an early date.

In Spain, a 'kill-site' at Torralba, near Gerona associated Acheulean hand-axes and remains of Straight-tusked elephants, which were no doubt hunted by their makers.

A famous Great Interglacial human skull comes from Steinheim, in Thuringia, again with a warm-forest fauna attributable to some part of this Interglacial.

Another of the very few known human relics of this period is a jaw-bone from the cave of Montmaurin, Ariège, in the extreme south of France. It is a curious fact, not hitherto satisfactorily explained, that hardly any European cave-deposits can be attributed to a date earlier than the Last Interglacial, but at Montmaurin, in a fissure (La Niche) near to the main cave, remains of a breccia (obviously a preserved trace of an earlier filling) of Great Interglacial age has in some way survived erosion. It yielded this jaw in association with a characteristic mammalian fauna.

In central Europe, Elster and Saale were the two greatest glaciations, which brought their ice-edges southwards to only slightly different maximum extents in different places. The Lower Older Loess found in the periglacial zone of the former and which covered the Mauer Sands containing the jaw of Heidelberg Man (Fig. 34), was completely weathered through to its base at this site during the Great Interglacial. This is clear proof of the very long period of temperate climate to which the soil-formation was due.

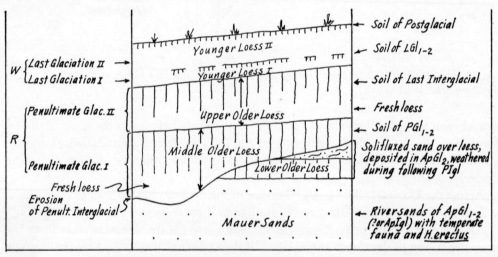

34. *Mauer, near Heidelberg: north face of brick-pit, showing succession of loesses and soils.*

That the Great Interglacial was not an uninterruptedly warm period is shown by the formation of two river-terraces in the valleys of the Rivers Ilm and Saale, in Saxony and northern Thuringia, between the boulder-clays of the Elster and of the maximum Saale glaciations. The second of these may reasonably be attributed to the first phase of Saale, which was nowhere as extensive as the second, but the former (Glacial Terrace I, as it is locally called) must fall well within the Great Interglacial and represent a recurrence of colder conditions favourable to climatic aggradation of continental rivers (Table, Fig. 28).

In the Rhine Valley, near Schaffhausen, the Main Terrace corresponds to Mindel and the High Terrace (also called 'Upper Middle') to Riss, but there is no representative, so far south, of the River Saale's Glacial Terrace I, which would fall between them. There is, however, a very marked deepening of the valley, due to the Great Interglacial, between the two glacial Terraces.

Geological evidence from important sites thus seems to point to the occurrence of at least one, if not more, distinctly cold intervals within the Great Interglacial, marked by interrupted aggradation and renewed river-downcutting in response to a temporarily lowered sea-level at Swanscombe and a minor (cold) climatic aggradation by rivers in central Germany. No glacial readvance at this time can anywhere be demonstrated, because the succeeding Riss/Saale main glaciation must have overridden and re-worked any of its deposits or morphological features.

CHAPTER 4

Outline history of the Pleistocene

Part II : Upper Pleistocene and Postglacial

The Upper Pleistocene, as defined by Woldstedt (1958), comprises the latter two main glaciations and the Last Interglacial which divides them (Table, Fig. 35).

The Penultimate (one-from-last) Glaciation, Riss of the Alps, Saale of northern Europe, Gipping in East Anglia and Illinoian of the United States, was also a double advance. The radial extent of Riss in eastern Europe did not quite equal that of Mindel II, but surpassed it almost everywhere in the west (Fig. 30). In the Alps, however, Riss II represented the maximum extent of the ice-front. It seems likely that the Interstadial between them was brief and not well marked, and so with a decidedly cool climate. Since the Last Glaciation was hardly anywhere as extensive as either phase of Riss/Saale, the boulder clays of the latter are still prominent in glaciated landscapes beyond the Last Glaciation moraines, as, for instance, in East Anglia ('Older Drift') and north Germany.

In northern Europe there is only one Saale glacial end-moraine betokening the maximum glaciation, but in the Alpine region there are *two* climatic terraces of the Rhine, the Upper High and Middle High Terraces, the latter of which corresponds, by stratigraphical links, with the local glacial maximum as marked by end-moraines of Riss. In Britain, we have *two* Gipping boulder-clays in East Anglia (see below).

There is, well inside the Saale moraines in north Germany, another set of relatively fresh end-moraines called Fläming, or Warthe. These are clearly subsequent to the Saale glaciation and have most recently been interpreted by Woldstedt as belonging to a later and lesser advance of Riss/Saale ice. Because of the relative freshness of their topography, however, as also because of the evidence of the Thuringian river-terraces (below), which already account for two advances of Saale, Zeuner assigned the Fläming moraines to the first phase of the Last Glaciation.

In Thuringia, two of the climatic terraces of the River Saale (Glacial Terraces 2 and 3) are clearly anterior in formation to the Saale ground-moraine. The former is covered conformably by an Older Loess, this being followed, unconformably, by Saale boulder-clay. The latter, near Halle, has deposits of Saale directly and conformably upon it. This strongly suggests that Terrace 2 corresponds to a first Saale cold period, in which ice never reached the district, but during which, in periglacial conditions, the terrace was buried under the contemporary loess. This last was eroded in an Interstadial, so that when the Saale ice eventually overrode it, the contact between loess and boulder-clay showed an unconformity. During the Saale advance to its maximum, Terrace 3 was formed before the ice reached it and was then covered conformably by the boulder-clay when at length it did so. This would prove two phases of Riss in Thuringia, of which the second (Saale) was the more extensive and so supports Zeuner's attribution of the Warthe end-moraines to the Last (=Würm) Glaciation, not to the Penultimate (=Riss).

In East Anglia there are, in places, two chalky boulder-clays, both later than that of the Lowestoft maximum glaciation. This is most clearly seen at High Lodge, near Mildenhall, Suffolk, for in many places they are not seen in superposition and so are not easily distinguishable as the products of distinct advance-phases.

At High Lodge, a depression in the lower of these held a small lake, in the deposits of which are preserved plant-remains, animal bones and the flint implements of early men who left their traces of occupation of the shores at three distinct successive horizons in the deposits of a Gipping I–II Interstadial, for the whole site was later overridden by the boulder-clay of the Gipping II ice-advance. One of the industries recovered includes the well-known High Lodge developed Clactonian side-scrapers.

Further south, between Hatfield and Watford, a blue chalky boulder-clay containing derived Jurassic fossils, probably that of Lowestoft (=Mindel II), lies 15 to 20 ft. below the present surface in a gravel pit at Sleapshyde, Herts. It is covered by a considerable thickness of flint outwash-gravels. Above these, near Bricket Wood, is another, brown, boulder-clay, now completely decalcified by subsequent weathering. This represents the extension of a lobe of the Gipping ice in this district which reached, at its maximum, as far south as Aldenham. Another similar ice-lobe filled the Finchley depression from the north-east. This was probably of Gipping II (=Riss II) which, like Saale in Germany, seems to have been the more extensive of the two advances.

In the Lower Thames, Gipping/Saale is represented by a phase of deep downcutting by the river to the bench (N-Bench of Zeuner, 1957) of the Taplow, or Fifty-foot, Terrace. This bench lies at only 3 ft. (1 m.) above present river-level at

THOUSAND (10³) YEARS B.P.

RADIATION DATES, ON LOGARITHMIC SCALE

	ALPS	N. EUROPE	POLLEN ZONES, LOESSES, SOILS
		Dates by C¹⁴	Pollen-zones

Scale	ALPS	N. EUROPE	POLLEN ZONES, LOESSES, SOILS
PRESENT	POSTGLACIAL		Subatlantic VIII
			Subboreal VIIb
		5	Atlantic VIIa
		6	Boreal { late VI / early V }
10		8	Preboreal IV

Dates by C¹⁴: V

Pollen-zones
- Subatlantic — VIII
- Subboreal — VIIb
- 5 — Atlantic — VIIa
- 6 — Boreal { late VI / early V }
- 8 — Preboreal IV

	RHINE	ILM-SAALE	N. EUROPE	Pollen zones / Loesses / Soils
20			Finiglacial retreat / Fennoscandian moraines	Younger Dryas III
			Gotiglacial retreat	10 Alleröd II
	III 25	Low-low Terrace / III Floodplain	Middle Swedish moraines / Daniglacial III retreat / Pomeranian II moraines / I	**LATE WÜRM** / Upper Older Dryas c / 12 Bölling (warmer) b I / Lower Older Dryas a / 15 Younger Loess III
30			W.II-III Interstadial (Masurian or Paudorf)	22·5 Paudorf soil on Y.L.II
40	LAST GLACIATION / 72 / II	WÜRM / Mid-low Terr. / II Terr.6	WEICHSEL / Frankfurt+Brandenburg moraines (Z.) or / Frankfurt-Posen only (W.)	**MIDDLE WÜRM** / Younger Loess II
50			W.I-II Interstadial (Rixdorf or Göttweig)	28 / 35 Göttweig soils 2 / 40 1
60				**EARLY WÜRM**
70	115 / High-low Terr. / I Terr.5	WEICHSEL (W.) / Brandenburg moraine or / WARTHE (Z.) / Fläming moraine	50? Younger Loess I / 70?
80	I			
90			Skaerumhede Series	Krems (Lr. Austria) soil and widespread deep weathering of Older Loesses and all preceding moraines and glacial deposits.
100	LAST INTER- GLACIAL	RISS-WÜRM / Terr.4	EEMIAN / Danish Middle Bed (cool)	
110				
120			Eem Series / Hötting Breccia (flint)	
130				
140	187 / Low-high Terr. / II Terr.3		Fläming moraine (W.) / 'JUNGRISS' or Saale II (Z.)	Upper Older Loess
150		RISS	SAALE	Interstadial soil
160	II			
170	PEN- ULTIMATE			
180	GLACIATION / 230 / High-high Terr. / Terr.2		Saale II (W.) or Saale I (Z.)	Middle Older Loess
190	I			
200				UPPER
210				
220				
230				MIDDLE

35. Correlation-table for Upper Pleistocene.

BALTIC STAGES, SEA-LEVELS		THAMES EUSTATIC TERRACES	G. BRITAIN GLACIAL FEATURES	N. AMERICA	
Baltic Sea	F L A N D R I A N	Modern Floodplain		Medithermal	P O S T G L A C I A L
Litorina Sea		Tilbury aggradation, with intervening pauses marked by peat-formations.		Altithermal (Hypsithermal)	
Ancylus Lake				Cochrane moraine	
Yoldia Sea				Anathermal	
Baltic Ice-lake			Scottish corrie-glaciers	Valders moraine	W I S C O N S I N
				Two Creeks Interstadial	
			Loch Lomond Readvance	Mankato Advance	
				Interstadial	
(slight regression)		3rd Buried Channel cut	Highland Readvance (West Lowland, Perth and Moray Firth moraines)	Cary advance	
				Interstadial	
				Tazewell advance	
Versilian transgression		Ponders End aggradation		Interstadial	
				Jowan advance	
				Interstadial	
-100 m.?		2nd Buried Channel cut	Scottish & Welsh Readvance (Aberdeen-Lammermuir, Lake-District and North Welsh moraines)	Farmdale advance	I O W A N
Epimonastirian + 2-4 m.		Lower Floodplain Terrace 4 m.		Interstadial	
-100 m.?		1st Buried Channel cut	Irish Sea-York-Hunstanton glaciation	Altonian advance	
Late, 6-8 m.		Upper Floodplain Terrace 7.5 m.	Barrington (Cambs), Ipswich (Stutton, Bobbits Hole) inter-glacial beds, Brundon (Suffolk)	SANGAMON INTERGLACIAL	
Monastirian		Argile rouge weathering (Somme)	Charing Cross & Selsey faunas		
Main, 15-18 m. (=Tyrrhenian II)		Taplow Terrace 15 m.	Upper Summertown-Radley Terrace (Oxon.)		
-200 m.?		Ebbsfleet Loess	Upper Chalky Boulder-clay GIPPING II Wolvercote Channel (Oxon.)	ILLINOIAN	
			High Lodge lake deposits		
↑		Main Coombe Rock (erosion to ± O.D.)	GIPPING I Great Chalky Boulder-clay at High Lodge & Hatfield		
PLEISTOCENE					
↓ PLEISTOCENE					

Dartford and is graded to a very low sea-level (perhaps — 200 m. o.d.), the coast-line at that time being far out in the North Sea basin.

In the west, a true boulder-clay at Fremington, near Barnstaple, Devon, contains Scottish erratics, showing that an Irish Sea ice-sheet, earlier than the lower (25-ft.) Last Interglacial shore-line, reached north Devon (Fig. 30). This probably represents Riss II of the continent, but it might be earlier.

On the slope of the valley-side at Swanscombe (Fig. 32), the Upper Gravels represent the solifluction and slope-washing of all glacial periods since the deposition of the Boyn Hill Terrace materials, beginning with the two Penultimate phases. At Baker's Hole and Ebbsfleet, both a little further to the east, the first phase of the Penultimate Glaciation is represented by the Coombe Rock, a valley-fill consisting of a mass of frost-fractured, slumped and solifluxed Chalk. This was in places cut away by the Ebbsfleet River, a small tributary of the Thames, during the Inter-stadial and, on the bare Chalk bench thus formed, was laid down a thick deposit of true loess, with typical calcareous concretions (Lösspüppchen) during the second glacial phase. This is one of the few typical loess deposits found in Britain (Fig. 36). All are in the south-eastern corner of the country, nearest the Continent and most are more or less water-derived and not *in situ*. Another, possibly of the same date as the Ebbsfleet Loess, is to be seen in a cliff-section at Pegwell Bay, near Ramsgate, Kent. These deposits point to a truly continental and periglacial climate in south-east England during this time.

Riss/Saale is represented in periglacial areas on the Continent by thick deposits of Older Loesses, so-called because of their often greatly weathered character and yellow-red colour, which distinguish them from the Younger Loesses, these often very fresh, in contrast, and of a pale yellow colour. They are attributed to the Last Glaciation (Würm/Weichsel).

At Mauer, near Heidelberg, we saw (Fig. 34 p. 74) that the Lower Older Loess and its overlying river sands were weathered completely through during the Great Inter-glacial and in part eroded away during the same Interglacial, following the weathering. On this eroded surface was laid down a fresh Middle Older Loess during Riss/Saale I. The upper part of this loess was weathered *in situ* to a loess-loam, a soil darker in colour and with clay increased at the expense of silt-grade materials, during the I–II Interstadial. In Riss/Saale II, more fresh Older Loess was laid down on the weathered soil. This, in its turn, was deeply weathered during the following Last Interglacial, before the Younger Loess series was deposited over it. Each weathering-horizon thus

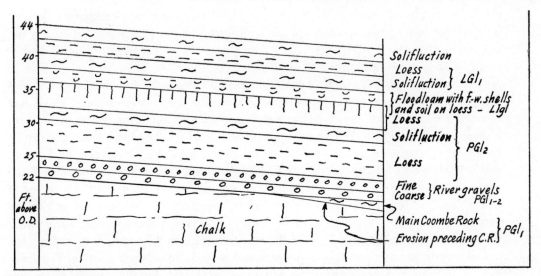

36. *Ebbsfleet Valley, near Swanscombe, Kent: British loesses, after Burchell.*

denotes a temperate interval between the phases of loess deposition and each fresh loess is the periglacial concomitant of a glacial advance outside the Middle Rhine area.

Quite apart from their characteristic contained pre-Mindel II fauna, it is this succession of loesses, and the weathering-loams formed on them, that gives a minimum relative geological date of Mindel I–II to the Mauer Sands which underlie them. The fauna, in fact, suggests a date one step (Günz-Mindel Interglacial) earlier yet, but the geological evidence is insufficient to prove this, though it may well be the case.

Older Loesses are found in many other famous sections in western, central and eastern Europe. It is universally agreed that they represent the Riss/Saale glaciation, or, when more than two are present, as here, even Mindel/Elster. Few individual exposures are as complete, or present such unequivocal conclusions as does Mauer. (Compare that at Sedlec, Fig. 26.)

Three Older Loesses occur also at Achenheim, Alsace. The earliest of these, the Lower Older Loess, probably belongs to the second phase of Mindel/Elster. It rests on fossil-bearing river sediments of the Rhine containing a fauna comparable with those of Mosbach and Mauer. The Middle Older Loess is atypical and is perhaps not

81

F

in situ, being humic and strongly weathered. It may represent materials derived from the Lower (higher up the slope of the Vosges), and here redeposited during the Great Interglacial. The Upper Older Loess contains a cold-steppe fauna and is clearly due to the second phase of Riss/Saale.

In France the Older Loesses appear in many sections, as in the famous gravel and brick-pits of the Somme, near Amiens and Abbeville, which have yielded the classical Abbevillian (Chellean) and Acheulean hand-axes (Commont, 1912, Breuil & Koslowski, 1931). From the often deep weathering which they have undergone since their formation, the loess-loams so formed are called '*argile rouge*' (red clay) in French.

We have seen that the Riss/Saale glaciation (and, in particular, its second advance-phase) was in many places a very intense glaciation, fully equal, in extent of its ice-sheet, to Mindel/Elster, and even over considerable distances across Europe exceeding the latter. Its effects in the contemporary periglacial areas of Europe were also intense. One of these, little noticed in the literature, is to be seen in cave-deposits.

Only rarely are any pre-Riss deposits found in caves in western Europe. The fragment of breccia from La Niche at Montmaurin, which yielded the jaw of a presumed Great Interglacial man, is one already mentioned and shows that, in many other cases, caves must originally have contained earlier fillings than those now to be found in them. Almost invariably, however, the deepest layers of cave-earth can be assigned to no earlier date than the Last Interglacial. In a few instances, no doubt (see the Grotte du Prince, below), it may be that the cave concerned was not in existence earlier, or was inaccessible to sedimentation, but this certainly does not apply to all.

The cave of La Mouthe, near Les Eyzies, for example, opens on the limestone plateau, perhaps 100 m. above the level of the present River Vézère. In this situation a water-worn cave system – or indeed a solution-cave – must be extremely ancient, for it was formed at a time when the local water-table in the porous limestone was at this level – i.e., before the present valley-system was cut. It must, therefore, be older by many tens of thousands of years than the earliest deposition of its present filling: a red cave-earth of Last Interglacial date, just like that to be found in many other caves in the region at much lower levels – and therefore by that much younger in date of formation.

It seems, therefore, that the periglacial cold of Riss/Saale, or its immediate sequel, the improvement of climate at the approach of the Riss-Würm Interglacial, with possible floods of meltwater coursing over frozen soil and subsoil rock, initiated processes of intense erosion which, in the vast majority of cases swept the caves completely of their pre-existing fillings, leaving clean limestone floors on which the sedimentation of the ensuing Interglacial could begin afresh. The glaciation which followed (Last Glaciation, Würm/Weichsel) seems not to have had any similar results, for sedimentation and accumulation of deposits of human occupation often continued uninterrruptedly in the caves from the Last Interglacial up to the present day.

The Last Interglacial is called Riss-Würm in the Alpine region, corresponding with the Eemian of north-western Europe, the Ipswichian of East Anglia and Sangamon in North America.

In comparison with the Great Interglacial it was only a short interval, estimated by Penck & Brückner as of three times the length of the Postglacial (this being taken as of 20,000 years' duration), i.e. 60,000 years. It was nevertheless fully temperate and indeed warmer at times than at present in Britain and the Low Countries, but included a cooler oscillation.

In the Alps, it is marked by valley-erosion, whereby the rivers cut down through the cold aggradations of the Lower High Terrace, corresponding with Riss II, to the bench of the highest of three Low Terraces. Considerable weathering took place on the surfaces of the moraines of all previous glacial advances. That on the Riss moraines, though resulting in marked soil-formation, did not penetrate to such depths as had the long-continued Great Interglacial weathering on those of Mindel. It was on this evidence that Penck & Brückner arrived at the estimate of the length of the Last Interglacial quoted above. Evidence for a colder spell within the Last Interglacial is lacking in the Alps.

In the valleys of the Ilm and Saale rivers in Thuringia there is, however, a Glacial Terrace 4, which, though lower than Terrace 3 (which, it may be recalled, was over-ridden by Saale boulder-clay near Halle), stands higher than two more Low Terraces and the present floodplain. It seems possible that Glacial Terrace 4 corresponds to a renewed cold-climate aggradation within the Last Interglacial (Table, Fig. 35).

At Rüdersdorf, near Berlin, a boring to a depth of close on 180 m. passed through three boulder-clays separated by two 'Interglacial' beds. The lower of the latter, the so-called 'Paludina Horizon' (named after a characteristic fresh-water snail), is close on 16 m. thick and perhaps represents a lake- or marsh-environment of the Last Interglacial. The upper (Rixdorf Horizon) consisted of gravels and sands only 5 m. thick and contains a cool or cold fauna, including mammoth. It appears in comparison to reflect only a minor ice-retreat.

In Denmark, a series of temperate-climate deposits (Herning Series) lies between evidences of two glaciations. The temperate layers were studied by pollen-analysis and the spectra found at the different horizons gave a detailed picture of local vegetation-changes.

A lowermost phase of subarctic flora, with tundra-plants, gives way to a first temperate to warm forest and fresh-water flora, the latter including *Brasenia purpurea*, a species of water-lily now extinct in Europe but surviving in southern China. This phase corresponds to a shallow marine transgression in the Low Countries, called the 'Eem Sea', which, locally, has given its name to the whole Last Interglacial. Renewed colder conditions next ensue, first with coniferous forest and then with tundra-plants again (Danish Middle Bed). This, in its turn, gives way to a second temperate forest and pond-flora with *Brasenia*, followed by increasing cold and finally a periglacial solifluction-deposit (Skaerumhede Series). Once again, but now more clearly still, for we are that much further north, near to the effects of even slight readvances of ice, we have evidence of a markedly cold phase subdividing a Last Interglacial which, in both earlier and later temperate incidents, shows a climate distinctly warmer than that of the present day.

In southern Britain, the Ipswichian Interglacial is named from a site at Bobbit's Hole, near Ipswich, which yielded lake-sediments with temperate plants overlying Gipping boulder-clay. The pollen-diagram shows one cold-warm-cold climatic oscillation and, though *Brasenia* and other typically warm-climate species are absent, the plant-succession and general picture are similar to those known from the Eemian of the Continent. The fresh-water deposits extend well below present sea-level, but it is likely that, so far east, they have been involved in the general North-Sea downwarping since their formation.

The Eemian Sea, at its maximum, stood at about 50 ft. (17·5 m.) higher than at present. Beaches at this level are known as 'Main Monastirian' in the Mediterranean. From this it fell during the cold interval, with a renewed rise to 25 ft. (7·5 m., Late Monastirian) in the latter part of the Last Interglacial.

These two positive movements of sea-level and their standstills at the maximum heights caused two eustatic aggradations in the Lower Thames, those of the Taplow (or '50-foot') Terrace and the Upper Floodplain ('25-foot') Terrace, the summits of which lie close to those heights.

At Trafalgar Square and Regent Street, London, recent deep excavations in Taplow gravels for the foundations of new buildings have yielded plentiful bones of a warm forest fauna, including hippopotamus, and, at the former, plant-remains typical of the Last Interglacial in Britain.

In the cliff-section of Black Rock, at the eastern end of Brighton promenade, there is a well-known exposure of the Late Monastirian beach, standing at 7·5 m. above present high water. This presents the ancient Chalk cliff, running back at some angle to the present-day coastline, with beach-gravels at its foot containing bones of large animals, including Straight-tusked elephant and horse. The whole is covered by 'head', or chalky solifluction-deposits, due to the freeze and thaw of the Last Glaciation. The modern cliff cuts through this valley-fill (Fig. 37).

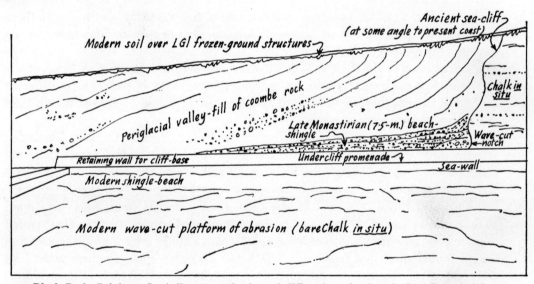

37. *Black Rock, Brighton. Semi-diagrammatic view of cliff-section, showing the Last Interglacial beach.*

This late Last-Interglacial sea-level at 7·5 m. is recognized at many places, in Devonshire and elsewhere, by beach-platforms and wave-cut notches in hard rocks which stand well above present sea-level.

The earlier, 17·5 m. beach was found at the base of thick cave-deposits at the Grotte du Prince, Mentone. The cave is a sea-cave cut at high-water mark by waves of the Main Monastirian Mediterranean. This beach, like other Interglacial beaches of the Mediterranean, contains a characteristic marine shell, *Strombus bubonius*, no longer present there but living today off the west African coast in Senegal. This indicates Interglacial water-temperatures distinctly higher than at present.

The Last Glaciation (Würm, Warthe+Weichsel, Wisconsin) was triple, as is generally agreed on the north European field evidence. There are three Low Terraces of the Rhine, three distinct deposits of Younger Loess in some parts of the periglacial zone and three phases of marine regression to be discerned on many coasts and in the lower valleys of some rivers. In addition, there are at least six fresh-looking glacial end-moraines in north Germany (Map, Fig. 30), of which the more southerly are interpreted as representing the maxima of three main successive advances of decreasing intensity, while the remainder are due to standstills or minor readvances during the retreat of the final phase. Evidences of two Interstadials divide the three principal advances, and there are several minor warmer oscillations between the pauses in the general retreat.

The ice-sheets of the Last Glaciation were nowhere as extensive as those of the two preceding major glaciations, being from 5 per cent. to 15 per cent. (in different areas) less in area than the maximum, both in the Alps and in the surroundings of the Scandinavian centre.

Opinions differ as to whether the first or the second phase of the Last Glaciation was the more intense. Both in north Germany and in the United States there are anomalous sets of moraines – the Fläming (Warthe) and the Iowan, respectively – which are assigned by some to a late phase of Riss/Illinoian ('Jungriss' of some Continental authors), but (notably by Zeuner, 1945 and Holmes, 1964) attributed to the earliest phase of the Last Glaciation. A somewhat temperate interval separated them respectively from Weichsel/Wisconsin, which Zeuner interprets as representing a First Interstadial (Last Glaciation I/II) while his opponents refer them to the Last (Riss/Würm) Interglacial and regard the outermost Weichsel (Brandenburg) moraines as due to the first phase of the Last (Weichsel) Glaciation. All the evidence to resolve this difference of opinion does not, so far as we yet know, exist, but two observations of Zeuner are strongly in favour of his interpretation: there are, as we saw, some good suggestions of two phases of Riss/Saale in Thuringia, so that we do not need to press the Fläming Moraine into service as its later phase; further, the Fläming Moraine looks remarkably fresh, both morphologically and in its state of chemical weathering, to have undergone the denudation and alteration inseparable from the passage of even a relatively short full Interglacial. Even more, granting the correctness of the attribution of Warthe (Fläming Moraine) to Last Glaciation I, this must represent the maximum extension of Last Glaciation ice.

The school of thought which considers the Warthe glaciation as 'Jungriss' regards what it calls 'Würm II' as the most intense phase of the Last Glaciation. Such differences of opinion seem to arise from the differing evidences of the distinguishable number and thicknesses of loess-sheets and weathering-horizons in eastern and western Europe. There were, it is agreed, several glacial oscillations, which certainly were expressed with differing emphasis in different places. It is the shifting about among them of well-known 'labels' – what we chose to call even agreed field evidences – that have different connotations for different authors. The adoption of new terms for old, well-known phenomena adds to the confusion, which seems to be more a matter of words than of conflicting field-evidence. Thus, four phases of the Last Glaciation are favoured by some workers instead of the classical three of Soergel. These are called 'Early Würm', 'Middle Würm (Early)', 'Middle Würm (Main)' and 'Late Würm'. The only new thing here, apart from the names, is that 'Late Würm' refers to some pauses or readvances of minor scale, all later than Younger Loess III (=Würm/Weichsel III) so that the first three stages in fact correspond to Soergel's original Würm I, II and III, whatever may be thought about their relative intensities and extents. Holmes (1964) regards Würm II and Würm III of Soergel (followed by Zeuner) as belonging together, so he labels them 'Würm IIa and IIb' – nothing but a change of the names, for he agrees with Zeuner's correlations!

At the maximum of the Last Glaciation, the sea-level retreated once more to something like 100 m. below the present height. From the summit of the Upper Floodplain Terrace (Late Monastirian sea-level) formed at the end of the Last Interglacial, the Thames cut down the floor of its valley into a deep gorge, the First Buried Channel, as it graded its bed to the low sea-level.

Whatever we may call the readvances, there are three, sometimes intersecting, channels, successively cut to low sea-levels and refilled at the following slight rise, now buried by the final rise to the present sea-level (the Flandrian – Postglacial – transgression).

Deep borings and drainage cuts in the marshes of the Lower Versilia, Italy, confirm the triple Last-Glaciation sequence of low sea-levels, recovery from which produced intervening successions of coastal-dune and freshwater-lagoon deposits.

The lowest sea-level recorded was at more than −90 m. and presumably represents Würm I and corresponds with the First Buried Channel of the Thames.

It is certain that the climate of the First Interstadial (Last Glaciation I–II) was fully temperate, at least up to about the latitude of northern France. The evidence from fossil pollen further north suggests more boreal conditions. A mature weathering-soil was formed on the surface of Younger Loess I in Lower Austria at Göttweig, which separates it from a second Younger Loess. The name 'Göttweig' is often used far away from Austria to distinguish the First Interstadial more generally. At Paudorf, another village nearby, a distinct, though less mature, soil was formed after the cessation of deposition of Younger Loess II. This, similarly, has often given its name to the Second Interstadial (Würm II–III).

The First Interstadial was notable in Europe as that in which Neanderthal Man, the maker of the Mousterian implements, began to be replaced by modern races, the Cromagnon varieties, bearing different, Upper Palaeolithic, equipment. During a full glaciation, the north European and the Alpine ice-sheets left only a relatively narrow corridor between them in central Europe and this, owing to the meteorological effects of the ice on either hand, was subject to an extreme continental type of climate, producing a cold, dry loess-steppe environment (Map, Fig. 17). Unglaciated south-western Europe, even at times of lowered sea-level, must have enjoyed rather more oceanic and less extreme conditions. Even so, reindeer migrated, at least seasonally, as far south as the Pyrenees, so that the climatic advantage during a full glaciation was only relative. The retreat of the Würm/Warthe-Weichsel ice during the First Interstadial restored more habitable conditions to much of central Europe, and especially so to the extreme south-west, allowing the flowering of the classical earlier Upper Palaeolithic cultures there, with their naturalistic animal-art.

During this time the sea-level returned to a height of between 2 and 4 m. above its present level (Epimonastirian sea-level of Zeuner) and caused an eustatic aggradation of river-valleys within this range, to a height slightly above that of the present floodplain – e.g. the Lower Floodplain Terrace of the Lower Thames. Beaches, and marine platforms without beach deposits, at the Epimonastirian level are recognizable round many coasts where there are hard rocks to preserve them (e.g. South Devon and the Channel Islands). In some cases, as at Pear-Tree Cove, near Start Point, the platform was buried in 'head' – solifluction deposits – now forming the upper part of the cliff, the denudation of which under present-day conditions is exposing the platform at just over 2 m. cut by the Epimonastirian sea in the hard schists below.

In Britain, the maximum ice-advance ('Younger Drift' as opposed to the 'Older Drift' of the much-weathered and denuded preceding glaciations) of the Last Glaciation extended across the Wash on the east coast as far as Hunstanton in Norfolk (Map, Fig. 30). In the Midlands, north and west of Great Britain the situation was very complex and is still not fully explained in detail. Ice, mainly from Scotland, moved southwards down the east coast as far as Lincolnshire and Norfolk, but was prevented from pressing far inland south of the Tees by the higher ground of the Cleveland Hills, the North Yorkshire Moors and the Yorkshire and Lincolnshire Wolds. A local ice-cap in the Lake District diverted much of the Scottish ice coming from the Cheviots and the Southern Uplands eastwards through the Tyne Gap, the Eden Valley and the heads of Airedale and Wharfedale, into the Vale of York, forming, at its terminus, the York-Escrick Moraine, its own ice contributing to the flow. The westward expansion from this centre joined the main Irish-Sea sheet, which pushed a great lobe south-eastwards across the Cheshire Plain and into the Midland Gap, as far as the Wrekin. Local Welsh ice, centred on Snowdonia, similarly competed for space with the Irish-Sea sheet and, between them, they covered most of Wales, save for the extreme west, where the ice-edge reached only a short way inland along the southern shore of Cardigan Bay. In Ireland, the Midland General Glaciation, at its maximum formed the Southern Irish End Moraine, leaving only the most southern one quarter of the island's area ice free, save for a local ice-cap in the mountains of Kerry.

An interesting accompaniment of the maximum Last Glaciation advance in the east was an ice-dammed glacial lake in the Vale of Pickering, Yorkshire. The offshore ice blocked the original eastward drainage of the Vale, between Filey and Scarborough and the Scottish ice, coming south-eastwards via the Stainmore Gap into the Vale of York prevented its escape westwards. A great lake, Glacial Lake Pickering (Fig. 38), built up, filling the Vale completely as it received meltwater from the Cleveland Hills, the North Yorkshire Moors and the overflow from a similar smaller lake occupying Eskdale, as well as from the surrounding ice. This eventually found an overflow southwards by Malton and cut a deep gorge past Kirkham Abbey to join the Humber drainage, a course which the modern Derwent still follows. Lake-sediments filling the former basin in the Vale of Pickering form the level peaty 'carrs', which make excellent farm-lands today.

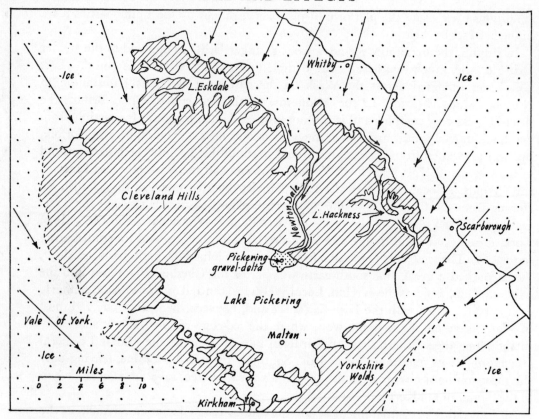

38. Last-Glaciation ice-dammed lakes and drainage in Yorkshire.

Following the retreat in the First Interstadial (Brörup in Denmark) there were several decreasing readvances of the main Scottish and the local Lake-District, Welsh and Irish ice-caps. The Scottish and Welsh Readvances, corresponding to Würm II, just crossed the Border, reached the northern tip of the Isle of Man and the extreme north of Antrim in Ireland (Map, Fig. 30). This stage is probably also to be correlated with the Aberdeen-Lammermuir end-moraines. In the Midlands, after the main Last Glaciation advance, various climatic oscillations have been described between sub-arctic and boreal, based on periglacial floras and insect faunas (Upton Warren Interstadial complex) and these have been dated by radio-carbon (see below) to between 50 and 20 thousand years B.P.

Following Last Glaciation II, corresponding to the Scottish Readvance in Britain and the cutting of the Second Buried Channel of the Thames to a sea-level perhaps still 100 m. below the present, somewhat milder conditions were established, called the Second Interstadial.

In Lower Austria and elsewhere, Younger Loess II began to develop a brown soil at Paudorf; in north Germany the Scandinavian ice retreated from the Brandenburg and Frankfurt-Posen Moraines for an unknown distance. In the valley of the Lea, a northern tributary of the lower Thames, an arctic plant-bed was forming (Ponders End stage) filling the Second Buried Channel. At Pin Hole Cave, in Derbyshire, unglaciated but not far from the margin of the maximum advance in the Cheshire Plain, both preceding cold stages had been marked by frost-weathered limestone rubble-layers. Upon the second of these the uppermost red cave-earth began to form and contained a poor Upper Palaeolithic (Creswellian) industry. On the Continent at this time the reindeer-hunters of the Magdalenian flourished in south-western France. It was a comparatively short respite.

Last Glaciation III followed, with spreading ice (Perth Readvance) over most of the Scottish Highlands, though the eastern coasts remained free. In the Baltic region, ice more once reached Pomerania and the Danish islands being represented there by three distinct end-moraines which suggest repeated oscillations between advance and retreat. These are probably reflected in a series of climatic oscillations frequently observed in periglacial deposits of Late Glacial age in north-west Europe (first in Denmark), which have been studied by pollen analysis (Figs. 30 and 35).

There are three stages of *Dryas* flora (an arctic assemblage including *Dryas octo-petala* (mountain avens) a tundra plant. These are called the Oldest and Older Dryas stages (Pollen Zone I) and the Younger Dryas (Zone III). Between the first two comes the Bölling Oscillation, with some pine and small-leaved deciduous trees – birch, willow, appearing, and this denotes a small, temporary improvement in the climate. After the renewal of arctic cold in the Older Dryas, there is a much more marked warming, with birch-trees dominant, called the Alleröd Oscillation (Zone II). This appears clearly in the British Isles also and, for instance, is even marked by a thin brown soil formed on chalky solifluction-fill in the Medway valley. In Ireland, it is a period of largely open heath-vegetation characterized by *Empetrum* (crowberry) and is associated with the Giant Irish Deer (*Megaceros*) which then flourished there for the last time before becoming extinct. The final onset of widespread cold is shown by a renewal of the *Dryas* flora in the Younger Dryas period (Zone III).

So ends the Last Glaciation. Thereafter the climate, as studied through the plant-pollen, improves steadily: through Preboreal (IV), Boreal (V and VI) to Atlantic (Zone VII). This denotes a maximum of mildness and year-round moisture favourable to the dominance of a mixed-oak-forest plant association, the so-called Postglacial Climatic Optimum. Only in central Scandinavia and the Scottish highland valleys can the last stages of glacial retreat be traced. That the Atlantic period did really represent an improvement beyond that of the present day may be seen in the more northerly distribution in Scandinavia of fossil remains of hazel of this time than is common today. *Trapa natans*, the water-chestnut, is found fossil in the Atlantic deposits of southern Sweden, while it extends no further than the southern shores of the Baltic nowadays. This is one of the species characteristic of the Eemian, with *Brasenia*, in the Netherlands.

The final shrinking and disappearance of the ice was accompanied by a rapid rise in sea-level, from perhaps — 70 m. or more to about its present level at the end of the Atlantic stage (Flandrian transgression). The bed of the North Sea, which had been largely dry land during the Boreal (as is shown by the occurrence of freshwater peats of that stage not uncommonly dredged up in fish-trawls) was now flooded once more by salt water. So, too, were the Straits of Dover and Britain became once again the island which it has remained ever since. This was certainly not for the first time, for, as Zeuner has pointed out (1946, p. 97), the presence of warm-water molluscs in the Eemian of the Netherlands suggests that the Channel was surely then open, for such animals cannot stand cold water and the only other way into the North Sea, about the north of Scotland, would have been impassable for them.

We speak of the present day as being 'Postglacial', from Pollen Zone IV onwards. Only minor changes of climate have taken place since then. Since the Climatic Optimum in Zone VII, there has been one slightly more continental oscillation (warmer summers, colder winters and somewhat drier conditions) in the Subboreal (Zone VIIb), with a reversion to milder, more cloudy, moister climate from a little before the Christian Era to the present day (Subatlantic, Zone VIII) (see Fig. 43, below).

Whether we are, today, in a full Interglacial (as we like to think!) – or merely another short Interstadial – it is perhaps too soon to say, but the Last Glaciation was not necessarily also Final!

CHAPTER 5

Supporting evidence for Pleistocene climatic sequences

Numerous branches of science have at different times and places played a part in building up the account of the Ice-Age events outlined in the last two Chapters. Of these, the biological sciences, botany and zoology, have, classically, been the most important contributors to the construction and interpretation of stratigraphical sequences. These have so much to tell us that they are separately treated below (Chap. 6). In contrast, some of the most recently-developed techniques, of nuclear physics among others, have revolutionized the whole matter of dating, and, here, too, it seems necessary to devote a special section (Chap. 8) to some of the currently-available methods. This Chapter, therefore, is concerned with other special branches of environmental research which help to interpret the Pleistocene field-evidence.

Geomorphology, the study of land-forms, lies partly in the field of geology and partly in that of geography. Nobody can understand the genesis of surface features of the earth without being acquainted with geological materials and processes, nor can a geographer accurately describe the present shape of the land without an understanding of how it came to be as he finds it. Until quite recently, both major sciences regarded geomorphology as really belonging to the other: an attitude symbolized by a German geologist's definition of it (*oberflächliche Erdbeschreibung*), which can be taken to mean either simply 'description of the Earth's surface' or 'superficial description of the Earth'. The latter derogatory meaning was evidently that implied!

Geomorphology is nowadays accorded recognition as a distinct speciality. Its essence lies in mapping and levelling in the field, as well as first observing the slightest features which may be worthy of study, and we have already seen many instances of its application to the explanation and ordering of glacial and periglacial phenomena such as the mapping of moraines and the levelling of river-profiles, sea-beaches and of the benches on which their deposits lie. Without this detailed work, it is impossible to read with any accuracy the successive events in the history of a countryside or coastline – and, even so, the absence of crucial criteria may leave us in some doubt as to the correct interpretation of what has been observed. Though it is the most recent Period of geology, the events even of the last few thousands of years have dissected and altered the surfaces and the exposed deposits of the Pleistocene, so that many pieces of the jigsaw-puzzle are missing and we may be forgiven if several equally valid explanations of what is left are often admissible.

Since we have few deep sections in Pleistocene deposits showing stratigraphical superpositions which would give a clear order of succession of geological and climatic events, the relative chronology of the Pleistocene is, in great part, supported on geomorphological foundations. It would be invidious to select names and works from among the many which have been published on the history of the Thames and tributaries, to take only one instance, but without the geomorphologists, the stratigraphers, palaeontologists and archaeologists concerned would have been hard put to it to explain and correlate their discoveries. Ten or a dozen distinct major stages of downcutting and aggradation have been described for the Thames valley, affording a well-established geomorphological and chronological framework into which new discoveries may be fitted in a logical order. It is not a final and complete order, and one which has gradually grown in detail and complexity since the beginning of the present century, because it has had to be modified to accommodate fresh information which would not fit acceptably into any of the existing divisions. This is a process which must continue, but we already have at least an outline of events.

An outstandingly important task still faces geomorphologists the world over: it is to study and correlate the evidences in their countries of ancient high (and low) sea-levels. Geological deposits, glaciations, animal fossils, plant remains, climatic evidences from soils and so on – all are fairly restricted to the areas in which they occur and so defy long-range correlations in time. Almost the only phenomenon which, by definition, may be expected to have had simultaneous worldwide effects of great diversity is the eustatically changing level from time to time of the Pleistocene seas. In the more closely studied parts of the world, such as Europe, the Mediterranean and the Atlantic coast of Morocco, a good deal has already been done, but the potential for inter-continental correlation of sea-level evidences has, as yet, scarcely been exploited elsewhere.

One most valuable piece of evidence relating to ancient sea-levels was pointed out by Zeuner (1945, p. 250) – that, if the average levels in stable areas of the high beaches and the duration of the respective Interglacials are plotted on the Milankovitch solar-radiation time-scale, a straight line can be drawn which intersects them all (with one minor exception) (Fig. 39). This line has a slope of 120 m./1 million years and indicates that, ignoring the wide, but relatively short-term, intervening oscillations of level due to glacial eustasy, the sea-level has been falling steadily at this rate throughout the Pleistocene. The cause of this is unknown, but has been attributed by Egyed (1956) to a slow expansion of the Earth as a whole (for which there is other evidence) and a consequent widening, and perhaps deepening, of the ocean basins.

If extrapolated backwards at this slope for 2 million years, this line also suggests reasonable dates for still older high marine platforms than those on which it was based, and these are supported by a number of dates gained by independent radiometric methods, in particular by Potassium/Argon (Fig. 39).

As for sea-beaches, the study of ancient high-level beaches of lakes in tropical countries may lead to elucidating a series of local events which may be relatable to past climatic changes in the drainage-basins concerned – high lake-levels corresponding to moister (pluvial), low water, or even subaerial weathering and erosion, to more arid periods. As with coastlines, reasonable assurance of tectonic stability of the region must be obtained before such land-forms may safely be attributed solely to climatic changes.

BOULDER-CLAY STUDIES

The original provenance of erratics found in a boulder-clay indicates the general direction of flow of the ice-sheet which transported them. The preferred direction of transport will probably lie within the angle subtended by the distant outcrop *in situ* at the point of collection of the erratics. Plots of erratics from different sources on a single orientation-diagram (like a compass-card) should support each other. It is a task for a competent petrologist to identify and trace to their sources any but the commonest and best-known erratics (West, 1968).

Striae cut in bedrock *in situ* by stones embedded in the sole of an ice-sheet moving over it will show by their preferred orientation the direction of travel of the ice at that place. Compass measurements of striae at a considerable number of sites in the

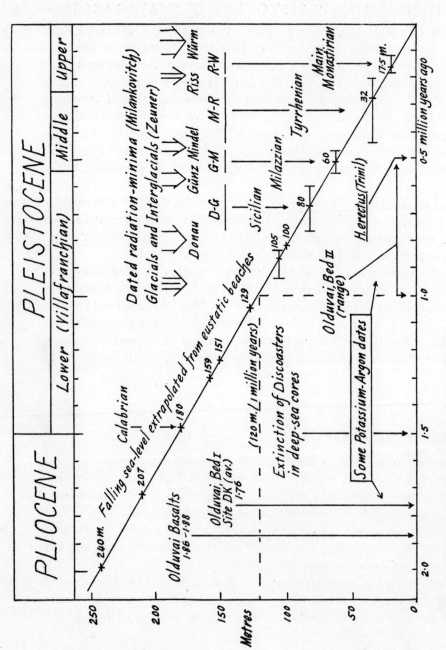

39. Height/time diagram showing secular fall in sea-level.

I. A valley-glacier. At the top left, the snowfield (névé, Firnfeld) of accumulation, lying in a cirque, from the edge of which the glacier tumbles in an ice-fall. The less steeply-sloping middle part is largely snow-covered, being still on, or near, the snowline. Another sudden increase in gradient, in the right half of the picture, leads to a second ice-fall before the terminus of the glacier. From the ice-front, now well below the snowline, issues a stream of water, braided on the narrow outwash-plain on the valley bottom.

To the right, in the middle distance, is a formerly-glaciated U-shaped tributary valley, with the slight 'V' of its present-day stream cut into the base of the 'U'.

Emergent peaks and all rock-surfaces above the levels of present and past glaciers show the sharp outlines and jagged pinnacles due to frost-weathering, in sharp contrast with the smoothed and worn glaciated surfaces. The very wide 'U' of the valley in the foreground shows that the glacier was formerly more extensive than it now is. Not only so, but, at some still earlier time, it also stood much higher, the latest abandoned narrow 'U' having been eroded through the bottom of a former, much wider, glaciated valley, just below the level of the 'shoulder' (bottom right) along which the road climbs into the side-valley.
Rhône Glacier, Switzerland.

II. *A valley-glacier with two prominent ice-falls. It is evidently winter, for snow lies thickly below the tree-line (foot of picture), so that surface-moraines and all but the major crevasses and irregularities are concealed. Here, as in Plate I, there is evidence of a formerly wider glaciated valley, extending up to the bare rock-shoulder on the left, above the snow-slope standing above the present glacier-level. All rock higher than this is intensely frost-weathered.* Glacier d'Argentière, Switzerland.

III. *Confluent glaciers. The frost-weathered rock-debris fallen from adjacent slopes forms lateral moraines at the margins of the ice. At the confluence, the two inner lateral moraines merge, to form a medial moraine. The existing medial moraines of the left-hand glacier show that it has already received one major and one minor tributary higher up its course.*

The freshly-gullied topography to the right of the glacier seems to have been recently cut, mainly by water-erosion, to judge by the V-sections, in what appears to be, not solid rock, but morainic material of a formerly much more extensive glaciation. Mount McKinley, Alaska.

IV. A valley-glacier, formerly much more extensive and massive (see the ice-worn rocks in the foreground). The rounded outlines of the now-emergent summits, in marked contrast with those of the preceding Swiss examples, show that, not long since, they, too, were completely covered by an ice-cap. A medial moraine comes from the tributary valley on the extreme left. Transverse and longitudinal crevasses show clearly, the former being convex in the direction of movement of the glacier, owing to friction with the rock walls, which retards marginal flow. Meltwater from the nose of the glacier begins the sorting of morainic material. Western Norway, unidentified.

V. Terminus of a glacier. Fissured and tottering masses of dirty ice, detached from the main body (top right) are thawing rapidly under a warm sun. Streams of meltwater (extreme right and to the left of the figures) form a pool (dark patch in the left bottom corner). Behind, on the left, rises a dark pile of bouldery end-moraine. Stones of all sizes lie at the feet of the figures (ground-moraine) and are being freed from the ice (englacial moraine) as it melts. Fox Glacier, South Island, New Zealand.

VI. *A formerly-glaciated valley. The ice-smoothed, U-shaped cross section is floored with thick morainic deposits. These have been cut into by streams, making steep-sided V-shaped gullies (left middle distance to right foreground), bridged or outflanked by the roads. Above the level of the glacier (where snow still lies in this picture), peaks and rock-surfaces are sharply frost-weathered.* Sedrun Tauetsch, Oberalpstock, Switzerland.

VII. *The head of a sea-loch (drowned glaciated valley). A late U-shaped valley has been eroded by a valley-glacier in a landscape of hard Cambrian rock, which had previously been planed off by an over-all ice-sheet of Last Glaciation age.* Roches moutonnées *cover this surface. Scree-slopes, due to Postglacial weathering (if not, in part at least, to lateral moraines of the valley-glacier) lie at the foot of the steep cliffs. The higher part of the valley, beyond the rock-barrier to the left, is occupied by a small freshwater loch, which fills a glacially-excavated rock-basin.* Loch Glendhu, W. Sutherland.

VIII. *A recently glaciated landscape of Tertiary basalt.* Roches moutonnées *in the foreground. Sea-lochs in the foreground and distance. Even Ben More, in the background (3,185 ft.) is ice-smoothed to its summit (hidden, here, in cloud). The cliffs across the water, with the enormous screes at their foot, are the result of Postglacial weathering of the closely-jointed lava.* Ross of Mull, Argyllshire.

IX. *A corrie which gave rise to a Late Glacial valley-glacier. The whole landscape, even to the summits, all close to 3,000 ft. in height, has been rounded off by an earlier ice-cap. The streams have failed to achieve much incision of the U-shaped valley floor even in the last 10,000 years, since deglaciation.* Wastdale Head, Cumberland.

X. *A U-shaped glacier-valley and the corrie at its head. The peaks are sharp and frost-weathered, and so were never overflowed by any later Pleistocene ice-sheet. Relatively much smaller glaciers would be expected in the Lowlands, this far south and west. The stream has begun to cut its 'V' into the base of the valley (most clearly seen in the immediate foreground).* Isle of Bute, Argyllshire.

XI. *The valley of a former glacier. The entire landscape is ice-planed. Note the U-shape of the valley, the truncated spurs and the intervening tributary valleys 'hanging', on the left. The river has attained a stage of some maturity by carving out a more or less level floodplain, over which it meanders from side to side, eroding on the outside bank at the bends and laying down sediments inside the curve. Thus, in time, it cuts into the bluffs on either side and redistributes the eroded material over the valley floor. Note, in the foreground and middle distance, the marks, in grass or crops, of several abandoned ancient meanders.* Glen Clova, Angus.

XII. *Drumlins forming islands in a sea-loch. A drowned glaciated lowland landscape.* Island More, Strangford Lough, Co. Down.

XIII. *Sandy boulder-clay* in situ: *the unsorted ground-moraine of a Younger Drift (Last Glaciation) ice-sheet, probably that originating in Snowdonia. The more rounded pebbles are of a relatively soft, shelly limestone. Note the preferred orientation of their long axes, from top left to bottom right.* Onibury, near Ludlow, Salop.

XIV. A glacial kame. The coarse,
unsorted glacial rubble, scarcely rounded
and in a clayey matrix, forms an
elevated mound or ridge, secondarily
slumped (note the stone-orientation) when
deprived of the lateral support of the ice
when it melted. Bradford Kame,
Lucker, Northumberland.

XV. Hummocks of drift (some partly
ploughed out) which are probably the
push-moraine (end-moraine disturbed by
subsequent ice-advance) of the ice-sheet
which laid down the Hunstanton
Boulder-Clay as its ground-moraine. This
is at the southernmost boundary of the
Younger Drift and is probably
attributable to the maximum advance in
the region of Last Glaciation I. Near
Weybourne, Norfolk.

region will be needed to indicate the general direction of the movement, because any one group, for local topographical reasons, may give anomalous indications. Arrows showing the preferred orientations at all the points studied may be plotted on a map and allow reliable conclusions as to directions of flow.

Where the bedrock is not exposed, or is too soft to have preserved striae, stones in a boulder-clay alone may yield evidence of ice-movement. They will tend to lie with their long axes parallel with the flow. Careful excavation of individual stones from an undisturbed exposure of till and measurements of the long-axis orientations of a hundred or so of them may be plotted as a rose-diagram to show the direction of ice-movement at that point. Diagrams from numbers of sites nearby should tend to confirm each other and justify the drawing of a general-direction arrow on the map (Fig. 40). It is usual, at the same time, to measure and plot the dip of stone long-axes (in a vertical plane), for if any particular favoured angle of dip is found, the deposit may not be *in situ*, have slumped or have been subject to movement since deposition, for instance by solifluction on a slope, and so be valueless for indicating the original direction of ice-flow. Well-confirmed directions for the movement of the ice of Older Drift glaciations in East Anglia have been obtained, using these methods (West, 1968).

VISUAL EXAMINATION

Not only the macroscopic characters of geological deposits are significant. In the study of sediments, whether formed by the action of water or wind, examination of the heavy-mineral grains under the microscope may be useful for tracing their provenance.

A standard method for the separation of a heavy-mineral fraction in a sediment from the preponderant bulk of quartz and other lighter grains is to shake a small sample in a closed tap-funnel with bromoform ($CHBr_3$). On standing, the light fraction floats, while all those grains denser than the liquid (specific gravity 2·89) sink and may be collected separately from below. Recognition and counting of the heavy components under the petrological microscope and their plotting as a 'spectrum' enable the heavy-mineral suites of different sediments to be compared and attributions to common, or distinct, sources sometimes to be made.

The coarser (0·5–1·0 mm) quartz-grains in a sediment, too, may yield useful indications (Cailleux, 1942). With oblique incident lighting under a low power of the

G

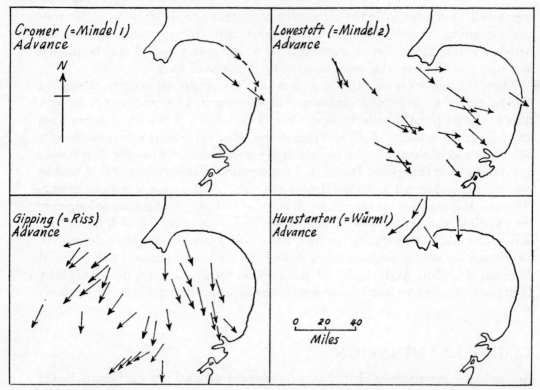

40. East Anglia: directions of ice movement, from preferred orientations of stones. After West (1963).

microscope they may be seen to be angular or more or less rounded; with highly-polished or matt surfaces. Angular grains and those with adhering ferruginous cement will have been comparatively recently released from their parent-rocks, while the more rounded and clean have a perhaps long history of transport by water or wind. Water-laid sediments will have a preponderance of well-polished quartzes; those repeatedly transported by wind, on the other hand, acquire not only well rounded forms but a characteristically matt or 'ground-glass' surface texture. The proportions of grains of the different classes in a sediment, obtained by counting and classifying at least 300 from each sample, will yield indications of its history. In zones now temperate, many round-matt grains in terrestrial deposits indicate formation in a periglacial environment. This may be most valuable when, as in the case of most quartzose sands, organic fossils are seldom preserved, so that they may contain little other evidence than that from the sand-grains themselves as to how or when they were formed.

Wind-rounded quartz grains are, of course, also found in desert sands. In sub-tropical and tropical zones, where there is no question of Pleistocene glaciation, well-rounded matt-surfaced grains are typical of former arid climatic phases in regions which, at the present day, have a good vegetation-cover which prevents wind-erosion and formation of aeolian sediments. Here, water-polished grains will be more frequent.

AIR-BORNE SEDIMENTS

Among the most valuable indications of climatic variations, often yielding detailed evidence of repeated changes, are sections, sometimes very deep ones, in loesses and the finer volcanic ashy sediments. Both having been transported and deposited by wind, they share the outstanding feature of being preferentially sorted into a rather narrow band of particle-sizes in the silt grade (0·06–0·002 mm.). The loess, as we have seen, has undergone this sorting during transport from the outwash-plains surrounding an ice-sheet to the place of its accumulation, perhaps hundreds of miles from its point of origin (Zeuner, 1945).

Volcanic explosions often throw up a cloud of solid or molten rock-material to very great heights. This consists of particles of all sizes, of which the larger, bombs and lapilli, fall at once fairly near the vent, while the sand- and silt-grade materials, caught by the prevailing winds, drift away to some distance before sinking to the ground. The silt, in particular, falls out only slowly, while anything finer yet (clay grade, <0·002 mm.) remains suspended in air-currents, perhaps for months. In the case of the dust-cloud from the Krakatao eruption of 1883, which initially reached a height of 80 km., the finest dust went right round the Earth and was detectable in the air for months afterwards. Ash-falls of silt-grade, like loess, may thus mantle whole countrysides relatively far from the vent. Cooled during their long passage through the air, they gently bury the pre-existing surface, without damaging it, so that, if the surface bore a weathering-soil, it may be preserved as a 'fossil' in its entirety.

The use of pyroclastic (particulate) volcanic products, especially lapilli, pumices

99

and volcanic ash-falls, as chronological markers for relative, and perhaps even absolute, chronology has been called 'Tephrachronology' (Greek: tephra=ashes). It yields absolute dates only if the volcanic materials themselves are qualitatively suitable, and old enough (more than, say 0·25 million years) to give Potassium/Argon dates. As stratigraphical marker-horizons, sometimes recognizable over large areas, they may be extremely useful. For instance, an ash, the product of an explosive eruption during the late Pleistocene from the Laacher See vent (Eifel district, West Germany) is widely distributed in the Younger Loess of the Rhineland and in peat-bogs (Frechen, 1952). As an isolated, dated event it enables confident correlations between any two sections in which it may appear.

In central Mexico, however, where there have been repeated similar (if generally lesser) explosions throughout the last several million years, coming from many different vents, it is first a question of distinguishing the material from among the many others like it and, if possible, assigning it to a particular vent and incident. This requires the intervention of skilled petrologists and vulcanologists to assist the stratigrapher, for the character and distribution of the product of a single explosion will differ from place to place according to height of the vent and the direction of the wind at the time of the explosion. Certain identification in particular cases is fairly hazardous in the presence of so many alternatives and such a wealth of very similar materials of different ages. A beginning has been made, but the results at the time of writing are not spectacular (Cornwall, 1968). A possible key to an extremely intricate problem has emerged by studying the several well-developed buried soils which denote longish pauses between eruptions of ashes and pumices. Unlike the volcanic materials themselves, the characters of these, mainly climatically determined, are fairly constant over much wider areas, so that particular past weathering-incidents may be identified and serve as stratigraphical markers among the volcanics, which vary from place to place. At certain locations in high mountain ranges, the volcanic sequence is interrupted by evidences of local glacial advances. In the tropics (Mexico City though high, 2,240 m., lies only 19° north of the Equator), advance of mountain glaciers depends more on increased precipitation than on any fall in average annual temperature. At present, rain falls only in summer, apart from isolated convectional storms (displuvial conditions), and the glaciers on the highest peaks of the Sierra Nevada are small. Two distinct series of buried reddish soils show, however, that in the not very remote past rainfall was much more evenly distributed throughout the year, and probably more plentiful (isopluvial conditions).

Between the two soil-forming periods there was at least one advance of mountain glaciers down to about the 3,000-m. level.

This is only as yet a fragment of a climatic sequence from more than thirty thousand years ago, uncorrelatable with any certainty with known events elsewhere, though it conceivably corresponds with the Last Interglacial of Europe, but it shows the potentiality of the method and points out a direction in which further research may lead to important advances in our knowledge. In Mexico, it is a potential method for fixing, in a dated volcanic and climatic context, the evidences of the at present earliest-known human inhabitants of the New World.

Silt-grade sediments are easily picked up and transported by water, so that, on any appreciable slope, fresh loess or volcanic ash is immediately seamed by erosion-runnels and gullies during heavy rain. Main drainage channels carrying considerable volumes of water in the wet season become deeply incised and the streams flow at the bottom of almost vertical-sided canyons. These are, in Mexico, often 200 m. or more deep, for it is a region of prolonged and intense volcanic activity and the great bulk of the superficial deposits consists of rather fine volcanic ash. In such sections the entire volcanic and climatic succession of at least the last several tens of thousands of years may be represented. Little work of interpretation has hitherto been done there, but, in Europe, loess-sections on a rather more modest scale have given us detailed evidence of Pleistocene glacial and interglacial alternations, the former represented by fresh deposits of loess, the latter by soils, more or less mature according to the conditions, formed at their exposed surfaces during milder intervals. The chemical and microscopical characters of these soils enable us to interpret the climatic conditions attending their formation. Few loess sections tell the whole story. Those of Mauer and Achenheim, already referred to, and that at Sedlec, in Moravia (Fig. 26), showing 8 loesses and 7 soils overlying a succession of terraces of the Vltava River, are among the most complete (Zeuner, 1945, 1964). The last, for instance, counting back from the Postglacial, must date back at least to the mild phase intervening between the two advances of Mindel. Those in the more oceanic West may show distinct weatherings betokening minor interstadial climatic oscillations, the effect of which never penetrated to the intensely continental climatic region of central Europe. It is for this reason that specialists in the two areas often disagree as to the detailed correlation of Pleistocene climatic sequences.

The characters of the buried soils in the European loess show that the climate of the Last Interglacial was, at least in part, somewhat warmer than that of the present day, that the First Interstadial of the Last Glaciation was fully temperate, though relatively short, and that of the Second Interstadial probably less warm than it is nowadays.

DEEP-SEA CORES

Oceanic deposits have independently contributed considerably to our knowledge of the climatic oscillations of the Ice Ages inferred from terrestrial geology. Oceanographers studying the sediments of deep-ocean floors have developed a piston-corer, lowered over the side of a ship, which is capable of taking and raising to the surface undisturbed cores of fine sediments up to 20 m. in length.

They represent slowly-accumulated deposits of mainly airborne mineral grains – terrestrial and meteoritic dust and volcanic ashes – mixed with the hard parts, siliceous and calcareous skeletons of surface-living drifting organisms (plankton), both plants and animals. When these die, their tiny shells and tests sink to the seafloor in their millions and are added, layer by layer, to the ooze gathering there. Changes in average surface-temperature of the water, such as will inevitably occur with widespread glaciation and deglaciation of the continents, are faithfully mirrored by changes in the species and proportions present of the plankton in the surface layers, and particularly of the Foraminifera, minute unicellular animals with calcareous tests of multifarious forms, corresponding to numerous different species. Sampling, recognition and counting of these fossils from each 10 cms. of a core will give a more or less complete history of changes in times past in the microfauna, and hence, of world climatic changes.

Numbers of cores taken in this way have been studied, from the Atlantic, the Pacific and the deep Mediterranean. Although there are sometimes breaks in the sequence, owing to slumping of unconsolidated sediments on submarine slopes during earthquakes and erosion by turbidity-currents, careful choice of coring-sites has enabled enough undisturbed series to be pieced together from different cores to represent almost the whole of the Pleistocene (Ericson & Wollin, 1964). Curves have been drawn (Fig. 41), based on the appearance and increase in numbers of cold-water species of Foraminifera at tropical sites, which greatly resemble that of Milankovitch representing computed changes in solar radiation received in high latitudes and it seems likely that we are here observing different results of a similar series of events.

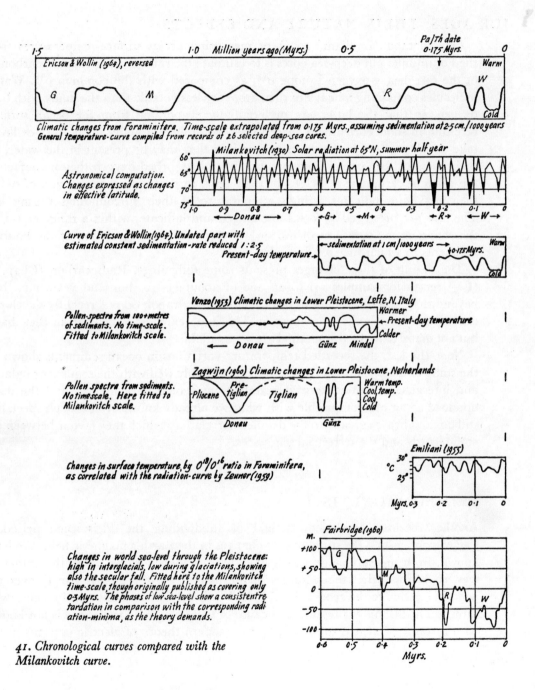

1·5   **1·0**   *Million years ago (Myrs.)*   **0·5**   *Pa/Th date* **0·175 Myrs.**   **0**

Ericson & Wollin (1964), reversed   Warm

G   M   R   W   Cold

Climatic changes from Foraminifera. Time-scale extrapolated from 0·175 Myrs., assuming sedimentation at 2·5 cm./1000 years General temperature-curve compiled from records of 26 selected deep-sea cores.

Milankovitch (1930) Solar radiation at 65°N, summer half year

Astronomical computation. Changes expressed as changes in effective latitude.

←—Donau—→   ←G→   ←M→   ←R→   ←W→

Curve of Ericson & Wollin (1964). Undated part with estimated constant sedimentation-rate reduced 1 : 2·5
Present-day temperature →

←—sedimentation at 1 cm/1000 years—→   Warm
→0·175 Myrs.
Cold

Venzo (1955) Climatic changes in Lower Pleistocene, Leffe, N. Italy

Pollen-spectra from 100+ metres of sediments. No time-scale. Fitted to Milankovitch scale.

Warmer
← Present-day temperature
Colder

←—— Donau ——→   Günz   Mindel

Zagwijn (1960) Climatic changes in Lower Pleistocene, Netherlands

Pollen spectra from sediments. No timescale. Here fitted to Milankovitch scale.

Pre-
Pliocene tiglian   Tiglian   Warm temp.
Cool, temp.
Cool
Cold

Donau   Günz

Changes in surface temperature, by O^{18}/O^{16} ratio in Foraminifera, as correlated with the radiation-curve by Zeuner (1959)

Emiliani (1955)

°C 30°
25°

Myrs. 0·3   0·2   0·1   0

Fairbridge (1960)

m.
+100   G
+50   M
0   R   W
−50
−100

Changes in world sea-level through the Pleistocene: high in interglacials, low during glaciations, showing also the secular fall. Fitted here to the Milankovitch time-scale, though originally published as covering only 0·3 Myrs. The phases of low sea-level show a consistent retardation in comparison with the corresponding radiation-minima, as the theory demands.

0·6   0·5   0·4   0·3   0·2   0·1   0
Myrs.

41. Chronological curves compared with the Milankovitch curve.

ICE AGES: THEIR NATURE AND EFFECTS

One method (Emiliani, 1955) of measuring ocean surface-temperatures from the Foraminifera in deep-sea cores is to estimate the ratio in their calcium carbonate of the rare heavy-oxygen isotope o^{18}, as compared with the common o^{16}. Water-molecules containing a heavy oxygen atom are less volatile than the many with light oxygen, so that the latter are preferentially evaporated from the ocean surface, resulting in some enrichment there in o^{18}. Foraminifera living in the surface layers take up o^{16} and o^{18} in whatever proportions they may be present in the water, for their chemical behaviour is identical. With the absorbed oxygen they manufacture the calcium carbonate ($CaCO_3$) of their tests, so that a high proportion of o^{18} in their make-up indicates a high temperature of their surroundings during life. The tests are preserved as fossils unchanged and indicate, within a range of 4–5°C, the changing temperatures of the surface during the accumulation of the bottom-sediments in which they are buried.

The dating of these changes presents some difficulties. Radiocarbon (Chap. 8) (C^{14}) serves for samples up to an age of about 30–40 thousand years B.P., but, beyond that, dates have had to be estimated by extrapolation. A recently-developed radioactivity method, Protoactinium/Thorium, extends somewhat further back, but not many dates by this method are yet available.

None the less, the recorded temperature-variations in ocean sediments shown by the study of deep-sea cores corresponds remarkably well with the sequence of glacial and interglacial periods derived mainly from geological evidences and the dates obtained so far are compatible with what we already know, even though specialists still do not always agree as to the detailed correlations which they favour between the core-records and the terrestrial sequences.

PALAEOMAGNETISM

Another recently developed method of subdividing the Pleistocene period is Palaeomagnetism – the study of past changes in the Earth's magnetic field. We have known, from measurements which have been made since the sixteenth century in western Europe, that the positions of the magnetic Poles change slowly over the years in relation to the geographical Poles by about 7′ of arc annually, in a clockwise direction in Britain at present. The cause of these magnetic variations is not known with certainty, but it is supposed, by the modern theory of electric currents in the

Earth's metallic core to be due to changes in direction of the dynamo-effect producing the Earth's magnetic field.

It has come as a complete surprise to us, only since the 1950's, that, in the past, the Earth's magnetic field has actually undergone complete reversal of polarity – and that on several occasions!

Past directions of the Earth's field are preserved in the remanent magnetism of igneous rocks, in which the grains of ferro-magnetic compounds, as they cool and solidify are lined up parallel with the Earth's field at their time and place. This inbuilt magnetism is preserved indefinitely, provided that the rock is not once more heated above a certain critical temperature (the Curie-point) at which the tiny oriented dipoles which maintain it are again mobilized. Rock-samples taken in the field, of which the present (and presumably original) orientation is carefully recorded, are examined in a zero ambient field by a sensitive magnetometer. This shows the declination (horizontal direction), dip (angle below the horizontal) and intensity of their ancient remanent magnetism. Dating of the crystallization of such rock-samples is by Potassium/Argon or other appropriate radiometric method.

When the magnetic orientations of several widely-separated contemporary samples are plotted on a map, the direction of the magnetic Poles at the time of their cooling is shown by the declination and, from their dips, their individual magnetic latitudes in their time can be determined.

It was by this means that the theory of Continental Drift, proposed by Wegener in 1915, at last, after 40 years, became 'respectable' and acceptable to geologists. (It should be interpolated here that, even if the continents have been, as the theory supposes, steadily drifting apart since the later Mesozoic, the rate of movement has been so slow that the Atlantic Rift has widened during the Pleistocene by only about 30 miles (48 km.). Thus, for studying the Ice Ages, the relative positions of the continents have remained essentially unchanged during the 2–3 million years of our interest.)

Palaeomagnetic study of many samples has now established that the present polarity of the Earth's field dates from about 0·7 m.y.a. and this is known as the 'Brunhes Normal Epoch'. Before that, there was a long (1·4 m.y. – 'Matuyama Epoch') phase of reversed magnetism, interrupted by three relatively short (average just over 0·1 m.y.) Normal Events. These were centred, respectively, on dates 0·95, 1·6 and 1·88 m.y. B.P. (Fig. 30, After Macdougall & Chamalaun, 1966). The earliest of them (Olduvai Normal Event, 0·16 m.y. long) is just a little older than Olduvai

Bed I, containing the earliest known Hominids, and so falls within the Late Villa-franchian part of the Lower Pleistocene. Earlier palaeomagnetic reversals do not concern our Period.

So far, few correlations have been made between known palaeomagnetic changes and the other Pleistocene geological, geomorphological, palaeontological and climatological events. There does not seem to be any causal connection between geomagnetism and any of them, but stratigraphical connections there certainly must be in areas where earlier Pleistocene volcanic rocks are in evidence to give palaeo-magnetic readings and dates to other deposits among which they have been laid down. The known reversals all antedate the main part of the Ice Age, save that the last reversed Epoch does overlap with the presumed dates of the Donau glacial phases (Fig. 28).

CHAPTER 6

Pleistocene floras and faunas

The ice-ages intruded, with repeated and cumulative climatic stresses, on a world which, for a very long time previously, had enjoyed widespread warm, equable oceanic climates. Their effects on living organisms, both plant and animal, were severe and in the case of many species (and even of larger groups) actually fatal. To many more, though surviving at present in parts of the world which afford them suitable conditions, their former wide distribution has been greatly restricted and dissected by Pleistocene geographical and environmental changes.

Climatic stress applied with the greatest intensity, of course, to what is now the North Temperate Zone, which at times of maximum glaciation was brought into what were virtually subarctic conditions; even the present-day subtropics and tropics did not escape changes which, if not involving catastrophic falls in average annual temperature, nevertheless altered the pre-existing character of their climates – as, for instance, in the direction of increased aridity, through southward shift of the desert belt. The net result has been a widespread impoverishment in variety of species of both floras and faunas since the Tertiary. This was accompanied by a much more marked zoning and localization of the distributions of sensitive species. Warm-climate plants formerly extended even to Greenland and Spitsbergen, in Lat. 81°N. At the same time it has favoured the spread and dominance in the affected areas of the less varied, but hardy, floras and faunas, which already possessed, or in the meantime acquired, adaptations to the severer conditions.

FLORA

The flora of the Eocene, where it has been preserved in the Lower London Tertiaries and London Clay, consisted wholly of plant-Families now exotic: for example

Robinia, the false acacia from the Americas, *Nipa* and other palms, cinnamon, figs, laurel and *Taxodium,* the swamp-cypress, among many others which survive today only in the Far East, Australia and the Americas.

In the succeeding Oligocene, the flora acquired a somewhat more temperate, though still warm-climate, character, with the date-palm, palmetto, witch-hazel and magnolias.

The brown-coal flora in western Germany, of Miocene date, was still subtropical, including *Sequoia* (the Californian big-tree genus), *Taxodium,* cinnamon, American and Asiatic species of walnuts, *Liquidambar,* the sweet gum, and, again, *Magnolia.*

In the Pliocene, the Reuverian (Netherlands) flora of north-western Europe included as well *Araucaria,* the monkey-puzzle, *Brasenia,* a far-Eastern water-lily, *Carya,* hickory, a North-American genus, and, in all, nearly half its numbers of eastern Asiatic and eastern North-American species.

In the Americas and the Far East there has been floral continuity since the Tertiary – only temporary redistribution there having taken place as the result of climatic changes.

In the mid-Tertiary, then, the flora was varied, but far more widespread over the globe than now. In Europe and the Mediterranean the less hardy members were gradually eliminated during the late Tertiary and Pleistocene. Advancing glaciations drove them southwards in all the northern continents, but, whereas, in eastern North America and in eastern Asia, there were no barriers to their retreat and subsequent re-colonization of lost ground, in Europe, western Asia and western North America mountains and seas cut them off and barred their re-entry, so that they have never recovered their former range, though, if now re-introduced and planted, many, like the magnolias, flourish out of doors in the north-western European climate of today, though they may not set seed.

Africa, noticeably, hardly comes into the picture as a refuge for the northern Old-World Tertiary flora. Not only is it practically separated from the rest of the Old World by the Mediterranean and Red Seas, but the waterless Saharan desert belt, broken in the north-south direction only by the narrow thread of the Nile Valley, constitutes an uncrossable barrier for plants.

Between the Late Pliocene (Reuverian) flora and that of Tegelen (Tiglian) come the Donau glacial phases of the Alps, no doubt with corresponding ice-accumulations in Scandinavia also, which, even if they never spread far south across the Baltic and north German plain, must nevertheless have caused much colder winters in the

Netherlands. This is to be seen in the Pre-Tiglian deposits, varying between 3 and 15 m. in thickness, which separate the Reuverian from the Tiglian temperate beds, and these have yielded a cold-climate flora. The something like 50 per cent. of exotic plant-species in the Reuverian was severely reduced to just over 10 per cent. in the Tiglian. *Tsuga* (hemlock-fir) and *Carya* (hickory), both North-American today, were among the last survivors. The rest of the trees were nearly all of modern European species.

Pollen-analysis, rather than macroscopic plant-fossils, has given most of the quantitative results of palaeobotany.

During the Reuverian, 90 per cent. or more of the plants represented were trees, forming an almost closed-canopy forest of extremely varied character, alder being specially prominent and the modern mixed-oak assemblage well represented. In the Pre-Tiglian, the exotic species disappear completely, being largely replaced by pine, with alder and others in subsidiary proportions, but at the maximum cooling of the climate more than 50 per cent. of the total flora was composed of herbs and grasses, latterly with a strong representation of low, woody shrubs like heather, crowberry and whortleberry – an open landscape of heath and moor with scattered woods, mainly of conifers.

Trees, including the few exotic survivors, returned to dominance in the Tiglian, but a representation of some 30 per cent. of the total by herbs, grasses and heath-plants points to a more open woodland than before, with pine and alder still the leading species, some spruce and small numbers only of broad-leaved deciduous trees.

Corresponding in Britain with the Dutch Tiglian is the lowest (Ludhamian) of three floral stages in the Red and Norwich Crags, studied by pollen-analysis (West, 1961) in bore-hole core-samples taken through the Crags at Ludham, Norfolk, from 30 ft. above to 165 ft. below present sea-level (Fig. 29).

These show a cool-temperate forest of a character similar to that of the Tiglian, including pine, spruce and hemlock. This gives way to a colder interval (Thurnian) with pine, alder and spruce, but a dominant heathland community. A short phase of open forest (Antian) is followed by a renewed increase to dominance of heath and grassland (Baventian). The summit of the Ludham core, after a sedimentary break, represents a new forest-zone, the Pastonian, and this, in another section, is followed by yet another cold stage, the Beestonian, preceding the Cromer Forest Bed, which belongs to a full Interglacial. The former two colder intervals are perhaps the East Anglian climatic equivalents of the two (or three?)

Günz glacial phases indicated on the Continent. The Pastonian and Beestonian are, so far, not represented elsewhere.

The climates represented by the oceanic-heath flora were little severer than that of the Highlands of Scotland today, but they are in marked contrast with the fully-temperate forest stages which, respectively, precede them.

Pollen-analysis has proved a most powerful tool for investigating floras of the past. The exines, or outer coats, of the tiny pollen-grains (the male gametes of flowering plants) have characteristic forms according to the genera of the plants producing them, and they may be recognized visually and counted under the microscope. In not too alkaline and well-aerated situations, moreover, they are wonderfully preserved for long periods of time. Some plants, among them many forest trees, release their pollen in uncountable myriads of grains on dry days. These are carried by wind and so may chance to settle on the female organs of one of their own species to fertilize them. The vast majority, however, blow away wastefully, eventually to settle wherever they are carried, and some of these are embodied in forming sediments, in soils and especially when they fall on standing water or on the moist surfaces of bogs or fens, they may be incorporated in deposits there being laid down and so be preserved.

This 'pollen-rain' is, over the year, representative of the vegetation of the region. After a sample of peat, a lake-bottom sediment or a buried soil has been suitably treated to remove mineral particles and amorphous humus, its pollen-content can be concentrated and mounted for microscopic examination. It may then give a faithful and quantitative account of the plant-species present (pollen 'spectrum') at the time of the sample's formation. A whole series of such samples in their stratigraphical order will reveal changes in the local pollen-spectrum with time and the results of grain-counts for the different genera and larger groups present may be plotted as a diagram to give a more or less complete history of floral changes at the site of the samples. Since climate is one of the main factors governing the type of vegetation to be found at a certain place or time, the pollen-diagram of vegetation-changes may be interpreted as a sequence of changing climates over the period of time represented by the samples.

We have just seen the evidences revealed by pollen-analysis in the periglacial zone of climatic changes accompanying advance and retreat of two major Lower Pleistocene glaciations – the Donau and Günz respectively. The record continues, during periods with vegetation, in certain well-studied parts of the world through the whole Middle and Upper Pleistocene and, beyond it, into the Postglacial period and right up to the present day.

During glaciations, the contemporary periglacial areas supported only arctic or boreal floras. When the ice-sheets retreated during Interglacials much richer plant-assemblages would successively develop as the climate improved, culminating, if the interval were long and climate mild enough, in a forest assemblage representing the climax association for those conditions. On readvance of the ice the flora, represented by its pollen, would again mirror deteriorating conditions by the changes to be seen in the species and concentrations of their products laid down at the time.

One subarctic peat is much like another, and the cold-climate floras which they contain are much the same for all glaciations. The most characteristic and extreme arctic plants belong to what is known as the 'Dryas flora', named from an arctic or alpine species, Dryas octopetala, mountain avens in English, Silberwurz in German, from the silvery sheen on the undersides of the fresh leaves.

With this are associated various dwarf trees of low, spreading habit. This is an adaptation to winter snow-cover, under which, the plant being dormant, it needs little or no light and air but profits by the protection afforded by the snow against extreme frost and wind-drying. Among them are several kinds of dwarf willow: Salix polaris, hastata, herbacea etc., most with small and frequently short, rounded leaves, in contrast with the long, relatively narrow leaves which we associate with temperate willow-trees. Here too are dwarf birch, Betula nana and other low woody growths, such as bearberry, Arctostaphylos, crowberry, Empetrum, again with numerous small, often rounded, leaves. So also are saxifrages and other plants growing in tight cushions. While some of them are still found today in high-mountain environments, not all are even alpine, for Salix polaris, Ranunculus hyperboreus, a buttercup, for instance, are known only from the treeless tundra of the far north. There were many grasses and sedges also.

This flora occupied the whole of ice-free Europe between the Scandinavian and Alpine ice-sheets during the glacial maxima, even in lowlands at distances of 300 km. or more from either ice-edge. The Ranunculus, for instance, has been reported from southern England.

With the Dryas flora are often associated marsh and water-plants: the common reed, Phragmites, marsh-marigold, Caltha, and semi-submerged and fully submerged species like pond-weeds. This is because, owing to permanently-frozen subsoil, the drainage of the lowland tundra zone is obstructed, so that many temporary pools of water stand on the surface through the summer. Further, the peaty sediments formed in these circumstances afford a favourable environment for the preservation of plant-remains and are therefore sought out by palaeobotanists studying ancient floras.

Not all available floras, of course, represent periods of maximum glaciation. During the advance and retreat stages of ice-sheets, the more extensive areas not yet, or no longer, ice-covered will present less severe climatic conditions and permit the development of floras with dominant coniferous or boreal tree-species, like pine, birch, willow and aspen scrub or forest. This will be marked in the pollen-diagrams by a strong increase in tree-pollen, as opposed to that of the herbs, grasses and low shrubs (non-tree pollen) typical of the tundra. In turn, the distinctly cold boreal flora will merge, to the southward, with more familiar broad-leaved deciduous-forest types wherever less severe conditions permit their development.

In contrast with the floras of glacial periods, those of full Interglacials had more individual character and they may even be identifiable by the presence, or increased proportions of particular species not shared by all. Though, in general, the succession of glaciations resulted in a progressive impoverishment of north-west European temperate floras, a few species now confined to the Continent did succeed in re-colonizing Britain, for example, during Interglacials. It must be remembered that a considerable number of tree-species, for instance, with which we are perfectly familiar in Britain today – spruce, sweet chestnut, walnut, to mention only a few – are the result of importation and planting in historic times, largely in the eighteenth century. Some have, by now, so established themselves and independently colonized wild areas far from the parks and gardens and plantations which they originally ornamented as rarities or economic imports that we tend to forget that they are foreigners and not members of the natural postglacial British flora. Scots pine (in the south of England), rhododendron, false acacia, respectively European, west Asiatic and American in origin, are among these.

Interglacial floras, especially of the earlier Interglacials, have been preserved, discovered and studied in only a limited number of places within the glaciated areas. This is quite naturally the case, because, in most places, a subsequent readvance of ice would plough up and destroy any plant-bearing deposits left in its course. Only if they were laid down in a lake-basin or river-valley and first comparatively deeply buried by, for instance, fresh outwash-deposits of the advancing ice, would any such survive intact. *Cromerian Interglacial.* The name-sites are on the coast of Norfolk, near Cromer, where the Cromer Forest Bed represents sediments of a considerable river or estuary which were covered to some depth by later deposits, before the advance over them of the first ice-sheet represented in Britain. Locally called the Lowestoft glaciation, it corresponds to Elster in northern Europe and Mindel of the Alps. As we have seen, earlier glaciations (Donau, Günz) known in the Alps show evidence of severe climatic cooling in Early Pleistocene northern Europe, but, here, there are no boulder-clays or other direct evidence of their extent, or even existence. The Cromer Forest Bed denotes the return to full Interglacial conditions following the cold phases described from the Crags.

West (1968) has codified the climatic and plant-successions of the three Interglacials and the Postglacial, as displayed in Britain, so that their distinct stages may be identified (Table, Fig. 42).

CLIMATE	VEGETATION-PHASE	ApIgl CROMERIAN	PIgl HOXNIAN	LIgl IPSWICHIAN	Pgl FLANDRIAN
cold	early-glacial	early Lowestoft	early Gipping	early Weichsel	—
cool	post-temperate	C IV	H IV	I IV	—
warm	late-temperate	C III	H III	I III	—
	early-temperate	C II	H II	I II	(Present day) F II (zones VII & VIII)
cool	pre-temperate	C I	H I	I I	F I (zones IV–VI)
cold	late-glacial	late Beestonian	late Lowestoft	late Gipping	late Weichsel (zones I–III)

42. Vegetation-phases of the latest Pleistocene Interglacials and of the Postglacial (After West 1963, slightly amended).

Each Interglacial begins with a Late Glacial phase marked by tundra or boreal-forest flora, with high non-arboreal pollen and no warmth-loving trees. This is

H

113

followed by a birch-pine forest, denoting a climate still severe, at least in winter, perhaps still distinctly continental. Then comes an Early Temperate flora of hazel, oak, ash, elm, lime, yew, holly – the mixed-oak forest – followed by a Late Temperate phase, in which hornbeam and sometimes spruce are added to the mixed-oak forest. The recurrence of glaciation is heralded by the Post-Temperate stage, in which the mixed-oak forest trees are replaced by pine, fir, birch and alder. This gives way, in its turn, to increasing grassland and heath, marked by a strong increase of non-arboreal pollen with few or no broad-leaved trees in the succeeding Early Glacial period.

With minor differences, this general pattern is confirmed for all the Interglacials, as far as they are known, and for the Postglacial.

The *Cromerian Interglacial* has a few species foreign to the modern British flora, but all of them are European – spruce, fir, water-fern (*Azolla*) and water-chestnut (*Trapa*) among others. For the rest, it is chiefly peculiar for the practical absence of *Hippophaë*, the sea-buckthorn, in the late-glacial (often common in other Interglacials), the unimportance of hazel at any stage and the marked preponderance of elm over lime in the early-temperate phase.

The Great Interglacial, or Hoxnian, of East Anglia begins with sea-buckthorn and heath, followed by birch-pine with grasses. The mixed-oak forest comes in next, with alder and hazel strongly represented. Fir is present in small amounts throughout. Hornbeam and spruce appear later. In the final stages, pine, birch and grasses increase, the alder being the last of the broad-leaved trees to disappear and crowberry comes in towards the end. *Acer*, the maple, is present in the temperate forest phases, but never in more than slight amounts.

In the *Last Interglacial*, Ipswichian, oak is preceded by elm in the temperate phase. Maple and alder are of almost equal importance. Hornbeam replaces maple in the late-temperate forest. Hazel is very prominent through both early- and late-temperate stages. Pine also maintains itself throughout. Spruce and fir, in contrast with the two preceding Interglacials, though present from the beginning of the forest phase, remain unimportant. Lime is scarcely represented at all. There are a few foreigners, but *Azolla* is absent, while *Brasenia*, a water-lily now confined to the Far East, is typical of the corresponding Eemian of the Netherlands, though it never, apparently, reached Britain.

The Postglacial floral history of many parts of central and western Europe is much better known and in greater detail than that of any of the Interglacials. Well preserved lake, marsh and valley sediments, from the west of Ireland to Sweden, from

Norway to the south of France, from Spain eastwards into central and eastern Europe, have been closely studied, mainly by pollen-analysis. Allowing for the great distances between the extreme sites and for local geographical factors, as well as for the progressive nature of the postglacial climatic improvement in a generally south to north direction, the diagrams given by stratigraphical series of samples show similar sequences of events. The samples, moreover, often consist of peats and sediments containing abundant organic matter, from which radiocarbon dates may be directly obtained, so that we now have a detailed chronology of the changing climates of the last 15,000 years in different parts of the Continent (Table, Fig. 43).

As in earlier Interglacials, a Late-Glacial stage, in which the *Dryas* flora is dominant, precedes the Postglacial proper but the eventual decrease of cold is heralded by two warmer oscillations, corresponding to rapid ice-retreats (Bölling and Alleröd) during which pine, birch and willow began to be prominent in the periglacial zone. This phase lasted for some 4,000 years (Pollen Zones I–III) during which there were two marked halts in glacial retreat, called the Older Dryas and Younger Dryas periods, during which subarctic conditions were continued.

At about 8,000 years B.P. begins the steady improvement of climate marking the beginning (Pollen Zone IV) of the Interglacial or Interstadial which we optimistically call the *Post*glacial! Birch is, as ever, the first immigrant tree. It has winged seeds which can blow long distances in strong winds and, with the favouring southerly gales which doubtless accompanied, in north-west Europe, the breakdown of the glacial anticyclone, it advanced rapidly. Pine followed in the wake of birch and in Zone V (Early Boreal) grew to dominance. Hazel next first appeared heralding the broad-leaved summer-green forest. The Later Boreal phase (Zone VI) saw hazel increase to its maximum and then gradually fall off again as it was overshadowed by elm, oak, alder and lime. Pine kept a steady subsidiary place, but fell almost to vanishing-point at the end of Zone VI.

Zone VIIa in the pollen-diagrams, called Atlantic, represents a change to a less continental, more oceanic type of climate, with less marked seasonal temperature changes, milder winters and warm, moist summers. This favoured an increase in alder, a moisture-loving tree, and the growth of both upland and lowland peat-bogs in the less well drained places, at the expense of forest. The transition from Boreal to Atlantic comprised what is known as the Postglacial Climatic Optimum, a time during which average annual temperatures rose to perhaps 2°C above those of the present day. The later part of the Atlantic period, beginning at about 5,000 B.P.

10³yrs. (C¹⁴)	CLIMATIC STAGES	POLLEN ZONES	FOREST DEVELOPMENT (S. Britain)	ARCHAEOLOGY
1 — A.D. B.C.	_Subatlantic_ Cooler, wetter, very oceanic, with mild winter	VIII — 1.5	Increase in birch continued. Lime and elm very low. Beech and hornbeam increase. Continued slight decrease in alder and oak	Historic
				Early Iron Age
1 — 2 — 3	_Subboreal_ Drier, more continental	VIIb — 3.0	Slight decrease in lime, alder, even oak; increase in birch. Decreasing hazel	Bronze Age
				Neolithic
4 — 5	_Atlantic_ Oceanic, warm, moist. (Climatic optimum)	VIIa — 5.3	Mixed oak forest, with oak and alder dominant, elm and lime. Hazel plentiful. Pine and birch are present, but unimportant.	Mesolithic
6 — 7	_Boreal_ Continental, drier, with increasing warmth	VI — 6.9	Pine decreasing, hazel maximum. Elm and oak appear and increase. Lime and alder appear at end.	
		V — 7.6	Pine dominant, birch decreasing. Hazel appears.	
8	_Preboreal_ Cold	IV — 8.3	Birch dominant, pine increasing	Late Upper Palaeolithic
9 — 10	_Late Glacial_ Subarctic to arctic	8.8 III	_Dryas_ flora. Rare willow & pine.	
		II — 10.0	Birch maximum	
		I	_Dryas_ flora, few trees	

43. _Postglacial forest-history and pollen-zones._

saw the arrival and spread in north-west Europe of the Neolithic cultures which, with increasing forest clearances for agriculture, represent man's first serious interference with the natural environment.

Zone VIIb, the Subboreal, corresponds to a return to slightly more continental conditions, with drier summers and colder winters. This is the time of the Bronze Age in north-west Europe.

Zone VIII is the Subatlantic period, standing for a recurrence, from the last few centuries B.C. up to the present day, of slightly cooler average conditions with abundant moisture brought by oceanic depressions. The natural vegetation has been progressively and increasingly modified by man and his domesticated animals, generally in the direction of clearance, exploitation and degradation by primitive agricultural, and latterly industrial, operations.

FAUNA

Animals are directly or indirectly dependent on plants and, since the character of the vegetation depends mainly on climate and soil, so the animals adapted to exploit that vegetation have distributions similar to those of their favoured food-plants. This applies to all animals alike, vertebrate and invertebrate, terrestrial and aquatic or marine, and all tend to evolve specializations that fit them to make the most of particular environments. Since, however, in the Pleistocene we are dealing with a relatively short period of time – a mere million or two years, as against tens, or hundreds, of millions for earlier geological periods – not all groups of animals show appreciable evolutionary changes in that time. It is above all the mammals that are important for the Pleistocene palaeontologist. This is not to say that other groups have been entirely ignored. Marine animals like Foraminifera and molluscs and terrestrial invertebrates like land- and freshwater-shells and insects have all contributed to our knowledge in certain circumstances. It remains true, however, that since the majority of Pleistocene deposits is terrestrial and continental, the land mammals, the fossil remains of which are likely to be found in such formations, provide the greater part of our information about Ice-Age faunas.

The heyday of the mammals was the Tertiary Era. By the Miocene they had reached their greatest individual size, numbers and variety of species and had spread widely in a world over the greater part of which climates and types of vegetation were favourable to their maintenance. With the Pliocene, a worldwide climatic cooling began to make itself felt and resulted in much-increased latitudinal contrasts in climate and vegetation and between coastal areas subject to oceanic effects and the centres of continents far removed from these. This process culminated in the repeated glaciations of the Pleistocene, during which the favourable warm-temperate zones became greatly circumscribed and for thousands of years at a time subarctic conditions extended to perhaps 40°N in the northern continents.

The mammals most dependent on temperate types of vegetation migrated southwards during glaciations and north again as the ice retreated and conditions improved. A few more adaptable groups in time diversified, evolving species with particular adaptations to less favourable environments, in occupying which they avoided for a time the pressure of competition from other species not so suited. Thus, animals of particular kinds were enabled to spread even into the very severe conditions of the arctic tundra and cold steppes, and these were able to hold their ground through glaciations, when their less specialized contemporaries were obliged to retreat.

As we have seen in relation to the plants, Europe and western Asia are divided by east-west mountain barriers, the Mediterranean and Red Seas and the subtropical desert belt from their southern continent, Africa. In the Americas, on the other hand, the Cordillera runs north-south, far to the west, and, in the more eastern parts of Asia, too, there are no serious barriers to plant and animal migration between higher latitudes and the tropical zone. In the western Old World, therefore, the Pleistocene glaciations took severe toll of sensitive and unadaptable species, both of plants and animals, while many such were able to survive in more southerly parts of the New World and the Far East: refuges from which they were later able to re-establish themselves in their former habitats when conditions improved. Those species which had become fitted to colder and drier environments, however, were enormously successful in their new habitats and, if the fauna as a whole was less varied as a result of Ice Age conditions, the well-adapted species increased hugely in numbers of individuals and provided rich hunting both for their carnivorous predators and for early men, who had similarly learned how to maintain themselves in cold climates.

Both the cold-adapted fauna of the Old World and the early human hunters found their way at a comparatively late stage of the Pleistocene into the Americas, via the Behring land-bridge which emerged on a wide front during a period of low eustatic sea-level. In the New World, as in the Old, the northern animals found plenty of space into which to expand and were present in large numbers during the late Pleistocene. The southern United States, Middle and South America retained representatives of Tertiary animal survivors (as did, and does, Africa) up to quite late times. A few species less conspicuous as sources of human food, such as the raccoon, opossum and armadillos are not uncommon, though persecuted, today. Many of the larger, mastodons, horses, camels, giant sloths and *Glyptodon*, an enormous armadillo, probably met extinction at the hands of early man long ago.

VILLAFRANCHIAN

The mammalian fauna of the European Lower Pleistocene, as represented at the Val d'Arno in northern Italy and Senèze in central France, is an unspecialized warm-climate assemblage with many Pliocene survivals. In this it is like the present-day African savannah fauna, which, similarly, has archaic survivals like rhinoceroses, giraffes and most varied antelopes, which have never been subject to the selective effects of repeated climatic stress. It is characterized by species almost all now extinct, though often belonging to genera which have modern representatives.

There is a monkey, *Macacus*, of the same genus as the surviving Barbary 'Ape' of Gibraltar and north Africa. Among rodents there is a porcupine, two extinct species of beaver and the voles (*Mimomys*) are of a primitive type with rooted, not continually-growing, molars. This last is a late specialization to counter the severe wear involved in cutting and chewing tough and dry vegetable food, which is all that is obtainable in more northern climes by their modern descendants.

The carnivores comprise two bears, of which one is the ancestor of the modern brown bear, the other of the now extinct cave-bear. Similarly, ancestors both of the living striped and spotted hyaenas are represented. There are six species of canids, all now extinct, the forerunners of foxes, wolves and jackals. The sabre-toothed cat, a large lion and an extinct species of lynx are found.

Proboscids are represented by the extinct mastodon, a primitive elephant (*Elephas planifrons*) and a more advanced species probably derived from it, *E. meridionalis*, the southern elephant. The typical rhinoceros is *Dicerorhinus etruscus* of which the modern descendant is the rare Sumatran rhino. There is a tapir, a genus now found only in South America and the Far East. Equids include a zebrine as well as two types of caballine horse, related to our modern tame horses, and, in Africa, at Olduvai, even the still more primitive three-toed horse, *Hipparion*, was a contemporary at this time of the early Hominids.

Among the ruminants are several extinct species of deer, known in the Pliocene, including an ancestor of the giant Irish deer, *Megaceros*, and a fallow-deer, *Dama*, but no red deer, elk or roe-deer as yet.

Cattle are represented by *Leptobos etruscus*, probably the ancestor of the later aurochs, *Bos primigenius*, and one of the modern bison. There is a gazelle, sheep or goat, a hippo and a wild pig.

EARLY GLACIATION (EG1), GÜNZ

The climatic cooling associated with the first Alpine glaciation caused many extinctions. The porcupine, mastodon, *E. planifrons*, the tapir, gazelles disappear, never to return. Other groups adapt and new species appear in the next temperate period. The Tiglian of the Netherlands (Günz I–II) shows *Trogontherium*, a giant beaver, immigrant from Asia, presumably pre-adapted there to more boreal or continental conditions. *E. meridionalis* persists, but alongside it appears *Elephas antiquus*, the well-known straight-tusked elephant, presumably derived from it, which long survives it in the succeeding Interglacials. A temperate-forest rhinoceros, *Dicerorhinus kirchbergensis*, first appears, probably a descendant of *etruscus*.

ANTEPENULTIMATE INTERGLACIAL (ApIgl), GÜNZ-MINDEL, CROMERIAN

The fauna of this Interglacial, as seen in the Cromer Forest Bed and in the High Terrace of the Somme at Abbeville, northern France, still includes a few of the old members – *Mimomys*, *E. meridionalis*, *Machairodus* (the sabre-tooth), an extinct lynx, *Leptobos etruscus* and *Trogontherium*. Two new bear-species appear and, for the first time, the modern wolf. *Elephas antiquus* comes in alongside *meridionalis* but there is no member of the mammoth lineage as yet. Both *Dicerorhinus etruscus* and *kirchbergensis* are present. Three new horse-species appear, all now extinct, with *Dama* and *Megaceros*, but beside these are found red deer, elk and roe, all of the modern species and *Bison priscus*, the now extinct species of European bison.

This is clearly a transition-stage, with modern species already fully adapted to the new conditions occurring alongside the last precarious survivors of the old régime, still persisting despite one glaciation passed.

ANTEPENULTIMATE GLACIATION (ApGl), MINDEL, ELSTER

The first phase of this was less severe, but still brought fatalities to both mice and elephants. In the Bacton Forest Bed, Norfolk, *Mimomys*, the primitive vole-genus, vanishes, giving place to *Arvicola*, the modern field-vole. *Elephas meridionalis* is doubtfully represented for the last time and a mammoth-like steppe-elephant first appears. The Bacton fauna still has a fairly temperate aspect, however, with the Etruscan rhinoceros, hippo and a variety of deer, with bears, hyaenas and other carnivores, but musk-ox and glutton, both nowadays inhabitants of subarctic lands, suggest lower winter temperatures than before.

The Interstadial, Mindel I–II (ApGl$_{1-2}$), is represented by the faunas of Mauer, near Heidelberg and Vértesszöllös, Hungary, at both of which are found relics of the earliest men so far known in Europe. They are still rich and varied faunas with a few, but striking, archaic survivors – *Trogontherium*, the big beaver, *Machairodus*, the sabre-tooth, *R. etruscus*, zebrine horses, *Leptobos*, the primitive ox – all making their last appearances before extinction.

In Mindel II they still appear at Süssenborn, near Weimar, east Germany, in deposits representing an early stage of the second glacial advance. Significant new arrivals at this time are *Rangifer*, reindeer, the steppe-elephant and rhinoceros, *E. trogontherii* and *D. hemitoechus*, and another cold-adapted immigrant from Asia, *Tichorhinus*, the woolly rhinoceros, unrelated to the *Dicerorhinus* genus.

PENULTIMATE INTERGLACIAL (PIgl) GREAT INTERGLACIAL, MINDEL-RISS, ELSTER-SAALE

Now a new faunal era begins. The Interglacial was very long, about 200,000 years, and faunas with minor differences, not by any means necessarily contemporary within that period, are known from several famous sites: Cannstatt, in south-west

Germany, both an early and a late stage from Swanscombe, Kent, Grays and Clacton in Essex. Gone are the Pliocene survivors. Extinct species there are, indeed, some, like *Dama clactonianus*, a fallow-deer, is characteristic of this period and of no other. A macaque monkey is known by a single molar tooth from Grays, Essex, and does not again occur in Britain. *Elephas antiquus*, with long straight tusks, is the elephant of forest environments; *trogontherii*, ancestor of the woolly mammoth of the steppes. A similar pair of temperate rhinoceroses are *Dicerorhinus kirchbergensis* (*merckii* of older authors) in forest, *D. hemitoechus* of open country. Brown (forest) and cave-bears are fully differentiated, as are the striped and spotted hyaenas (*Hyaena*, *Crocuta*). There is a heavily built forest-horse (*Equus germanicus*) and a light steppe-horse (*E. mosbachensis*). Oddly enough, long as was this Interglacial and evidently at least in part warm enough, *Hippopotamus* is not found, though it appears plentifully in the next. The modern wild forest-pig, *Sus scrofa* first appears. Beside the fallow deer are red deer, roe, the giant deer, *Megaceros*, *Bison* and the only recently (eighteenth century) extinct aurochs, *Bos primigenius*.

PENULTIMATE GLACIATION (PGl), RISS, SAALE

Early stages of both phases of Riss are represented by faunas from river-terraces in Thuringia. The first of them, from Glacial Terrace 2 of the Saale River, has a mixed temperate to cool aspect, with cave-bear, beaver and forest horse (*germanicus*). Both forest and woolly rhinoceroses are represented, large Bovines (*Bos* and *Bison* indistinguishable from the material) and an *antiquus* elephant tending towards the related steppe form, *trogontherii*.

The Saale/Gipping I–II (PGl $_{1-2}$) Interstadial is clearly displayed in a datable context at High Lodge, Suffolk. The mammals are not well represented among the finds, but the plant evidence is of distinctly boreal conditions with predominantly coniferous forest, and it is therefore thought to have been only a short, and distinctly cold Interstadial.

Evidence of the time of the second advance, Saale 2, comes again from central Germany, near Leipzig and Weimar. It is a poor fauna, but by now markedly cold-specialized: horse, woolly rhino, reindeer, musk-ox, *Elephas trogontherii* and true *E. primigenius*, mammoth, the long-haired northern species, appearing for the first time.

We here see the well-adapted glacial type of fauna which represents the cold-climate phases for the rest of the Pleistocene. Since the Interglacials represent, on the whole, a continuation of pre-glacial conditions, their faunas, taking into account the losses inflicted by the earlier glaciations, have, in comparison, a conservative aspect, with long-established temperate forest species. These evidently contrived to find southern refuge-areas during glaciations and to recolonize north-western Europe whenever the conditions there returned to 'normal'.

LAST INTERGLACIAL (LIgl), RISS-WÜRM

Despite the repeated stresses of the two very intense foregoing glaciations (Mindel and Riss), the Last Interglacial shows in several places an extremely rich temperate fauna, locally associated with the remains of the forerunners of Neanderthal Man. Among the sites are Ehringsdorf, near Weimar, Brundon in Suffolk, Trafalgar Square in London, Stutton, Suffolk and Peterborough, Hunts. A long list of carnivores (two bears, wolves, foxes, badger, otter, marten, wild cat, cave-lion, lynx, spotted hyaena) preyed on the varied gnawers, browsers and grazers. Among these were numbers of rodents, horses, forest-rhinoceros, pig, red deer, roe, fallow deer, elk, giant deer, wild cattle, bison, and (in the mountains) ibex and chamois. The straight-tusked forest elephant and hippopotamus are both so numerous and so large in size that they were evidently entirely at home in these conditions. Short as it was, in comparison with the preceding Great Interglacial, the Last, as the flora, too, shows, was markedly warm and genial. In London and near Ipswich, as in France, a water-tortoise (*Emys orbicularis*) is regularly found in Last Interglacial pond deposits. This, alone, indicates winter temperatures far less severe than those which we experience today.

LAST GLACIATION (LGl), WÜRM, WEICHSEL

The three or more phases of advance eroded still further the varied forest fauna of the Last Interglacial. With the first phase of the Last Glaciation, the loess-steppe/boreal-forest fauna established in the Penultimate Glaciation returned – cave-bear, arctic fox, cave lion, voles and marmots, steppe-horse (*E. przewalskii*), woolly rhino, reindeer, bison, mammoth.

The First Interstadial Würm I–II was a short but fully temperate interval. The evidence of sea-levels and soils formed on Younger Loess I at Göttweig and elsewhere supports this conclusion, as does a study of the fauna from the Upper Travertine of Ehringsdorf, for instance. A list of temperate rodents was preyed on by water-loving and woodland carnivores such as otter and polecat. Larger mammals were horses and half-asses, both the steppe (*hemitoechus*) and woolly rhinoceroses, giant deer, red deer, roe, reindeer, aurochs and bison. The mammoth was the only surviving elephant.

The second phase of the Last Glaciation, as displayed by the Younger Loess II all over western Europe contains the typical cold loess-steppe fauna. There are wolf, foxes, both red and arctic, ermine, polecat, cave lion and hyaena. Hares, lemming and marmots, even the desert jerboa, voles and hamsters formed their prey. Horse, woolly rhino, giant deer, reindeer, musk-ox and mammoth are the characteristic larger forms of cold or cool environments. At one stage even the typically steppe-dwelling saiga antelope (*Saiga tartarica*) reached the Thames at Twickenham, proving the extension at that time of a marked continental-type climate so far into the usually oceanic west.

Last Glaciation II–III, (LGl$_{2-3}$) the Second Interstadial (=Paudorf in Austria) provides only cold, subarctic faunas on the Continent and was clearly neither so long nor so temperate an interval as the First. The cave-hyaena was now extinct. The musk-ox withdrew to the north, but the lion, glutton, brown bear and the modern carnivores were much in evidence. The horse, pig, red deer, roe, reindeer were common. Ibex and chamois sometimes came down into the lowlands still.

Last Glaciation III was the last major advance of the Ice Ages. The cave-bear, the woolly rhino and the mammoth were almost the last of the familiar glacial fauna to become extinct at this stage. Other animals were mostly modern. The very last Pleistocene survivor was *Megaceros*, the giant deer, which, in Ireland, persisted up to the Alleröd warmer oscillation of the very end of the Late Glacial.

Mammalian survivors into the Postglacial include all the modern fauna, plus a few species, like aurochs, horses and latterly the European bison, which are now extinct, or verging on extinction, as a result of the destruction or invasion of their natural habitats by man. The brown bear, the elk and the wild pig persist still locally and rather precariously in remote eastern and northern swamps and forests of Europe.

To what extent human hunters contributed to the extinction of, for instance, the cave-bear, lion, giant Irish deer, woolly rhino or mammoth is unknown, but it seems in all these cases less likely to have been entirely natural, for these species seem long to have survived the severest natural hazards that the Ice Ages could muster, right up to the end of the Last Glaciation, and one would have supposed that the oncoming Postglacial presented no new ecological problems for them – save only, as suggested, in the increasing numbers and activities of man. Man has certainly been responsible for many extinctions in historic times and many surviving species are, today, endangered.

CHAPTER 7

Man in the Ice Age

The origin of the human species remains unknown in detail, though, by study of his physical, biochemical and psychological being in the light of modern discovery and by comparing him in these respects with his nearest living relations, the higher Primates, we now know enough of him as a member of this zoological group to have some shrewd ideas about how and when the Family of man separated from that of the great apes.

Two things about Primates in general are very striking: first that they are physically primitive and generalized; second, that they are mostly arboreal (tree-living) in habit. The first feature gave them physical plasticity – a potentiality to adapt themselves readily to changing circumstances – while the second endowed them with great agility, keen binocular vision in colour and a tendency often to adopt an erect posture, instead of the horizontal carriage of the trunk habitually used by all terrestrial quadrupeds.

Further, most Primates, and the great apes in particular, are tropical, indeed largely equatorial, in distribution and there can be no doubt, therefore, that the step in evolution from an ape-like creature to one more man-like took place in the tropics. If certainly originally a tropical forest dweller, it is astonishing how man has thence spread into every sort of natural environment and to every part of the continents, many of the places in which he lives successfully having the most severe and unpromising conditions of climate and environment for any living creature. One has only to think of the subarctic barren regions of the northern continents, on the one hand, and the hot, waterless deserts of the subtropics, on the other, to see to what extreme – and one would suppose uncongenial – conditions he has contrived to adapt himself happily.

Life, for a skilled climber among the trees of a tropical forest, is simple and pleasant, with vegetable food available in plenty at all seasons for the picking, no violent seasonal changes in temperature to cause discomfort and safety from most predators in the mere fact of being off the ground. There is, in fact, no internal stimulus at all to change a way of life so easy and satisfactory.

It is supposed that the necessity to change and evolve, which separated the man-Family (Hominids) from that of the apes (Pongids), came from without, in the shape of some climatic change which, over a period of thousands of years, caused tropical forests to retreat and wither, so that they could no longer support so large a population of apes as hitherto. Those apes living near the margins of the shrinking forest would find themselves obliged either to retreat from the zone affected or, if remaining, to adapt their habits to a changing environment. The climatic change concerned probably depended on the progressive cooling in the temperate zones which we have seen to have taken place throughout the Tertiary Era. This resulted in more marked zoning of the Earth's climates and the shrinking of the more widespread areas which formerly existed of equable, oceanic conditions. In the tropics, it may have resulted in a southward march of the desert and savannah belts at the expense of the equatorial forest.

Whatever the cause, the effect on man's ape-like ancestors was to oblige them to forsake the dwindling trees and become more terrestrial in habit, to collect items of animal food to supplement the increasing scarcity of fruits and succulent vegetable growths on which they had used to rely. In the savannah, they were greatly at risk from large carnivorous beasts and would have to become more wary, cunning in avoidance of danger, where their relatively puny strength, lack of natural weapons and modest fleetness of foot were handicaps to survival. These conditions would favour the bipedal erect stance, the flat-soled foot, the freed hands skilled in the management of inanimate objects, the growth of memory, reasoning and foresight and the development, we may suppose, of means of communication – language, indeed – in all of which postures, skills and faculties man far surpasses the apes and all of which were necessary to survival in the changed conditions of his life on the ground.

The transitional stages are obscure and scarcely illustrated at all by fossils, but the end-product is now tolerably well known, through discoveries, in south and east Africa, of remains of undoubted primitive Hominids, called Australopithecines. These, by the development of their skulls, teeth and limb-bones, with such interpretation of their contemporary environment as has been possible by study of associated natural evidences, can be seen to have been living in the savannah in precisely the way suggested above. They were gatherers, catchers and collectors (one can scarcely call them 'hunters' as yet) of animals for food, small and slow enough to be easily captured and killed. Some at least used sticks and stones, bones and horns, as weapons and implements, not at all, or scarcely, adapted at first in any special way for the purpose in view. Later, the first deliberately shaped implements appear – pebbles roughly shaped to a cutting edge by striking off flakes with intentional blows made with another stone. This stage was reached by *Homo habilis*, in Bed I at Olduvai, Tanzanía, in the Villafranchian, something like 1·75 million years ago.

By about 1 million years ago, in Bed II at Olduvai, there was a more advanced Hominid, now called *Homo erectus leakyi* (originally named by Leakey, his discoverer, the 'Chellean Man'), and he made quite skilfully-worked stone hand-axes in the Chellean manner. With these, and more probably with long wooden spears made by their use, he was hunting large game animals – hippo, large Bovids, Giraffids and antelopes – probably in organized parties, which, after a kill, settled down for the next few days to eat the quarry where it lay. Though the hand-axe does not seem to represent so tremendous an advance over the pebble-chopper as to have required three-quarters of a million years for its development, not only the technological improvement but those involving the strategy and necessary social cooperation for success in exploiting the larger and more dangerous animals for food, do represent a considerable step forward in man's increasing mastery of his environment, by first organizing himself. Knowledge grows at compound interest out of the pre-existing stock of knowledge and it is a very striking fact of Prehistory how painfully slow were the earlier steps, while, within the lifetimes of many of us today, the internal-combustion engine, radio communications, nuclear power, computers and space-flight, for instance, have developed out of the steam-era, from their most modest beginnings to the highly-developed technologies that they already are today. The rising curve of increasing knowledge, which was hardly perceptibly above the horizontal in the time of *Homo erectus*, begins, on the same time-scale, in our day, to be indistinguishable from the vertical!

After the lapse of another half million years, i.e. by about 500,000 years B.P., distinct races of *H. erectus* had spread out of Africa, as far as extreme south-east Asia (*H. erectus erectus* of Java, formerly called *Pithecanthropus*) and only a little later as far north as Peking (*H. e. pekinensis*) and even into central Europe (*H. e. ungaricus*), at Vértesszöllös, near Budapest, Hungary.

This represents not only a surprising radiation of a young species, across – indeed, probably, throughout – the whole breadth of the Old World in the tropical zone, all within a mere half-million years, but an enormous extension of its field also in latitude, from just south of the Equator in east Africa to Lat. 47°30′N. in Europe and from 12°30′N. in Java to 40°N. at Peking, in eastern Asia. Ecologically, it stands for an ability to adapt to a wide variety of environments, lying between the 70°F January isotherm in the south to the 20°F January isotherm both in Asia and Europe. This is an advance indeed! The Peking men and those of Vértesszöllös both used fire. The former were living in caves.

Vértesszöllös *H. erectus* is dated to the intra-Mindel Interstadial (Mindel I–II) of about 450,000 years B.P. The winter isotherms referred to above are those of the present day – not of a glacial period! This primitive species of man was by then already capable of surviving continental winters in the temperate zone. Succeeding glacial advances, however, were too much for him. We find another such at Mauer, near Heidelberg (*H. e. heidelbergensis*) at about the same time, if not, as is possible, even in the preceding full Interglacial (Günz-Mindel). Mauer is some 2° of latitude further north than Vértesszöllös and has a temperate-forest fauna, while that of the latter is distinctly cool in character, but with the advance of the Mindel 2 ice-sheets, the central European temperate fauna gives way to that of the loess-steppe. Man, with his pebble tools and his hand-axes, has not yet learned to make his way in such conditions and is forced to migrate southwards with the retreating type of environment to which he was accustomed.

It was not a great journey: probably not even a conscious migration for any one individual. A major ice-advance took some millennia to develop, and, at maximum, the nearest ice-front to Heidelberg was more than 300 km. away to the north-east, in Thuringia. As the forest thinned and shrank, the first Europeans retreated with it. The mountains and the seas may, at length, have entrapped those who could find no way over or round them, but a few – it needed but *very* few – somehow contrived a crossing or even the long march necessary to outflank the obstacles.

We find remains of evidently related men at Ternifine, Algeria, near Casablanca, Morocco, *H. erectus mauretanicus*. There they were living in temperate-moist environments quite unlike those places today, probably during the spread of Mindel-Elster glaciers in Europe, one climatic effect of which, perhaps was to produce a 'pluvial' period in what is now the desert zone. Their bones are found with those of

I

fishes, crocodiles and mammals like the hippopotamus, which haunted the waters of lakes and rivers now long vanished.

The Great Interglacial (Mindel-Riss, Elster-Saale) restored suitable conditions for man in Europe, but in spite of its great length, we have extremely little fossil evidence of the newcomers – three isolated specimens, indeed – although the flint hand-axes of Middle Acheulean types, made and used by them as general-purpose tools, are numbered in their tens of thousands, collected from numerous sites of this age. The European distribution of the implements extends from central Germany in the east through the whole of Lowland Britain to the west; north-south from the Rhine to Gibraltar and even further east in the mild Mediterranean environment, through Italy, as far as the Balkans and Greece. The hand-axes are found, of course, throughout Africa, the Near East and as far as India, but their makers do not seem to have penetrated to the Far East, where the comparable industries belong to the older chopper-tradition. There is no positive evidence on the point, but it seems possible that the descendants of *Homo erectus*, with a characteristically archaic industry, per-sisted longer here than anywhere else, just as did the archaic flora and fauna, relatively untouched by the events of the northern Ice Ages and their climatic repercussions elsewhere.

The two principal fossil skulls in Europe, those of Steinheim, near Stuttgart, and of Swanscombe, Kent, are of men (or rather of women, for both are considered to be female) of considerably larger brain-capacity than the various known specimens of *H. erectus*. Steinheim is by far the more complete, having well preserved facial and frontal bones, while that of Swanscombe consists only of the back and two sides of the brain-case, lacking the frontal and facial parts.

In form and measurements the two are very alike, as far as they can be compared, but the fact remains that, in the case of Swanscombe, only the unusual thickness of the bones distinguishes it from a skull of modern man, whereas the face and forehead of Steinheim are more allied, in some of their features, with Neanderthal man of the Last Glaciation, and the specimen is generally considered to represent a group ancestral to these. The third European example of this period is an isolated jaw from the cave of Montmaurin, Ariège, south-western France. It also has primitive Neanderthaloid characteristics. It remains uncertain, therefore, whether the Middle Acheulian people of the Great Interglacial in Europe were of a single type, or of two, or more! It is clear, nevertheless, that the two we know from their skulls both represent an evolutionary advance on *H. erectus*, in point of brain-size at least.

They lived, as the distribution of the flint implements shows, almost exclusively by the sea, or by rivers and lake-sides, apparently avoiding the uplands in between, and evidently hunted the large game animals at their drinking-places, the bones of which are found in plenty in the ancient river-gravels alongside the hand-axes and flakes which were presumably used to butcher them. The faunas are almost all of temperate-forest type and we can reconstruct their environment with some accuracy from the associated animal and plant remains.

Even the long Great Interglacial came to an end at last and once more the ice-sheets crept down into the lowlands from Scandinavia and the Alps in the two stages of the Penultimate (Riss/Saale/Gipping) Glaciation. Forest-haunting hunters and gatherers, still for the most part unable to adapt their ways of getting a living to the severer climate, were once more driven southwards, out of northern and central Europe. As before, some may have survived in the more genial climates of the peninsulas of Iberia, Italy and Greece, but even there the environment was not as favourable. Perhaps, again, north Africa was the main human refuge area, from which Europe was later to be re-peopled. We have the jaw of a primitive Neanderthaloid from Rabat in Morocco, which may be as early in date as the Riss Glaciation and the very low sea-levels of the time may have narrowed the Straits of Gibraltar and that between Sicily and Tunisia sufficiently to make the crossings practicable under stress.

The not very favourable inter-Riss Interstadial (Gipping I–II) at High Lodge, Suffolk, nevertheless proves that some hardy humans could push so far north even in the brief retreat between the main glacial advances, so that it seems probable that force of circumstances had taught some groups how to adapt to conditions less easy than those of a full Interglacial, so that retreat to far Africa had not after all been necessary for their survival.

Again the ice rolled back over their camping-places and, so far, we have not a single bone of these people to tell us what they were like in their persons. Only the imperishable flints which they lost or discarded prove that they were, without doubt, there at the time.

Penultimate Glaciation 2 eventually retreated and now Europe flowered again in the Last Interglacial; if for a shorter while than during the Great Interglacial, at least as generously to mankind. Human fossils, if still not numerous for this period, are widespread and varied. Most of the known specimens have a circum-Mediterranean distribution – in Spain, Italy, Palestine and north Africa, for instance, denoting perhaps a gradual spread from the glacial-stage refuges of the species, but a few pioneers made their way into central and eastern Europe – Taubach and Ehringsdorf, both near Weimar; Ganovče, in Czechoslovakia; Krapina in Croatia and Kiik Koba in the Crimea. Though differing considerably in anatomical detail, they nearly all have a family likeness in showing some features which became more pronounced in the widespread Neanderthal race of men during the Last Glaciation. They are generally regarded, therefore, as the direct forbears of the classic Neanderthal type. Only one Last Interglacial specimen seems not to conform to the rule – that of Fontéchevade, in the Charente Département of western France. This was a skull, like that of Swanscombe unfortunately lacking the crucial facial parts, yet with most of the important frontal bone preserved, and this seems to be of entirely modern *H. sapiens*, not Neanderthal, type. Remembering Great Interglacial Swanscombe, this Fontéchevade man leads to the suspicion that perhaps the direct ancestors of *H. sapiens* were already distinct, in the extreme west, from those of *H. neanderthalensis*, as early as the later Middle Pleistocene. Alas, the question cannot clearly be resolved as yet, for the theory of a pre-*sapiens* as well as a pre-Neanderthal type rests, in Europe, on these two finds alone, both defective as to the face, which might have proved or disproved the case. The idea is not without some support from outside Europe, for, in east Africa, at Kanam, Leakey found evidence of the existence there of a *sapiens*-like man dating from the Middle Pleistocene. Most palaeoanthropologists, however, seem to support the simpler view that modern man may have developed quite lately, in the Last Interglacial, or even since, from a Neanderthaloid ancestor rather like the well-known later Palestine specimens.

The Last-Interglacial Europeans, judging from the plant and animal remains found in association with their implements, enjoyed a distinctly warmer temperate climate than that of the present day – the Fontéchevade example, for instance came from a cave deposit which contained remains of the water-tortoise, *Emys orbicularis*, which could never stand the modern winters of north-western Europe. Though a number of sites in Britain has yielded rich Last-Interglacial floras and faunas, proving that the conditions here of that time were perfectly suitable to contemporary man, no single human specimen of the Last Interglacial has been found here and even flint implements securely datable to the period are scarcely known. It may be that the sea-level rose rapidly in the English Channel, cutting Britain off from the Continent before man had time to reach it, though the large Interglacial forest mammals had succeeded in doing so and flourished here in great numbers, perhaps untroubled by at least one of their usual predators!

At least one phase of the Last Glaciation (whether we call it Würm/Weichsel I or II – for authorities are divided) once more almost cut Europe in half. The ice-sheets of the Alps and of Scandinavia left only a narrow corridor of loess-steppe and tundra between them. Classic Neanderthal man of the extreme type had by now, however, developed physical and technological adaptations to cold conditions and was able to live, if not comfortably, at least successfully surviving the worst climatic stresses. Sheltering in caves and under suitable overhanging rocks, using fire, doubtless dressing in the skins of the animals which he killed, he stayed on throughout the earliest part of the Last Glaciation in Europe. His remains, and the characteristic Mousterian flint implements which he made, are found most plentifully in south-western France, but also in north Africa, Spain, Italy, Greece, the Near East and even as far into Asia as Usbekistan in the U.S.S.R. It is likely that they also occur widely in northern and eastern Asia, for the blank on the map is probably due to the fact that, over most of the area concerned, their remains have not been sought. Examples are unknown in Britain, but have been found as far north and west as Jersey, Belgium, the Rhineland and through Switzerland to central Europe.

These people were of a very characteristic physical type – short and stout in build, with often capacious skulls containing brains at least as large as those of many modern men. The skulls, nevertheless, were of peculiar shape – extremely long and low in the vault, the back of the head angulated to accommodate very powerful neck-muscles, the forehead over the eyes provided with an enormous and prominent brow-ridge. Their faces were long, the eye-sockets high and round, the nose broad, the jaws and teeth large, the chin receding. These are all primitive features, in their degree of development more ape-like than those of their immediate Last-Interglacial predecessors, so that when they were first recognized as early men they were naturally regarded as morphological 'missing links' between apes and men and the direct ancestors of the modern human species. With the discovery of Swanscombe and Fontéchevade, both so *sapiens*-like, yet earlier in date than the classic Neanderthal race, the whole Neanderthaloid line was thought to represent an extinct side-branch of the human stem, not in the main line of modern man's ancestry. Now, however, opinion is swinging back towards regarding them as possible ancestors, in view of the serious deficiencies in the only two Pre-*sapiens* examples and the impossibility, therefore, of proving the existence of an unequivocal *sapiens* type before the arrival in Europe of the men of the Upper Palaeolithic.

Neanderthal man, in the French caves, was a notable hunter of large game, mainly, it is supposed, with the spear, armed perhaps with the rather short triangular 'points' which form one of the types of the Mousterian industries. Whether he used such a weapon for throwing is doubtful, for even a short and relatively stout stone point is brittle and liable to be broken whenever it misses the target and strikes another stone in falling. It is possible that he used the rawhide noose, perhaps thrown as a lasso, and the *bolas* as a missile weapon. The occurrence of remains of considerable numbers of birds and smaller animals in the cave-deposits suggests that he was also a trapper, though nothing is directly known of his devices in this direction. Many of his characteristic implements – side-scrapers, for instance – were probably used for skinning and butchering carcasses and for currying skins, to make them soft and usable as wraps and garments for protection against the prevailing cold. It is sometimes forgotten, however, that, in summer at least, fairly high temperatures must have been experienced even at glacial maximum, for south-western France lies in a latitude quite different from that of Lapland, of which the present-day conditions are commonly compared with those of glacial western Europe. The winters, however, must have provided a severe test of human adaptation to cold.

With the First Interstadial (Last Glaciation 1–2) came an improvement in climate and with it, through the widening and increasingly temperate east-west corridor between the ice-fronts to north and south, incursions of men of the modern type (*Homo sapiens sapiens*), bearing new habits and new equipment. It has been suggested that, in France at least, there is evidence of evolution from the Mousterian industries of Neanderthal man to the Upper Palaeolithic blade-and-burin forms of flint tools. Ethnically, at any rate, there is an abrupt conclusion to the appearance of Neanderthal man and his replacement by men of the fully modern type, so that it looks as if, in geological terms of time-measurement at least, the former became suddenly extinct. The cause for such extinction is unknown – whether the two races fought and the Neanderthalers were slaughtered or whether they were peacefully absorbed, or, alternatively, merely economically outclassed, so that they declined and perished naturally in an environment in which, now, their special adaptations to cold conferred no ecological advantage. The outcome, in any case, is clear: they were everywhere quickly replaced by the newcomers and nowhere survived as a recognizable type beyond this time.

Upper Palaeolithic peoples of slightly varying *sapiens* type henceforth occupied the

Old World, from south Africa to the northern ice-front, from Spain to the south-Russian steppes and, in the north at least, their characteristic implements are sporadically known as far as northern China and eastern Siberia. South-east Asia, including the Indian sub-continent, has no Upper Palaeolithic industries to show and the reason for this is unknown. Perhaps an older culture survived here until late times, but the absence of any human fossil in India, for example, does not permit any guess at all as to the type of man inhabiting the area during the Late Pleistocene. The Far East has, indeed, yielded the Solo skulls, in Java, of a race with some Neanderthaloid characteristics, but their date is very uncertain. A primitive Australoid type of *H. sapiens sapiens*, like that represented by the Wadjak (Java) finds may eventually have populated Australia – but again we know nothing as yet about the possible date of such an event.

The origins of the European Upper Palaeolithic immigrants remain unknown. In western Europe, at any rate, they seem mainly to have come in from the east, and this would suggest western Asia, or, indeed, Africa, as their former homeland, in the latter case with Sinai, Palestine and Anatolia providing the likely route of access. There are some north-African affinities between both the stone industries and the cave-paintings of eastern Spain, and it is quite possible that there were some later Upper Palaeolithic arrivals across the Strait of Gibraltar from Morocco.

In the First Interstadial of the Last Glaciation they found in south-west Europe an ideal environment with a rich fauna, and this they not only hunted but pictured on their cave walls, engraved upon bone and ivory and carved in the round as decorations to their implements and as figurines.

The Cromagnon people of western France were a fine, upstanding race, the men often over six feet tall, the women of more average stature. Variants in other parts of Europe were not so distinctive, but all alike shared the skilled flint blade-work and the occasional excursion into artistic expression which is technically surprisingly accomplished, even to a modern eye. Neither before nor since has anything like it been produced in ancient times and the reason may well be in the comparative ease of living for a semi-sedentary hunter in their time, when the valleys were filled with an enormous richness of game and the mere business of getting meat did not fully occupy their time or their minds.

A recurrence of glaciation in the second advance of Würm/Weichsel brought this 'Golden Age' to a close. Once more the climate in western Europe became sub-arctic and the Magdalenian descendants of the first Upper Palaeolithic peoples had to adapt their ways to hunting reindeer instead of red deer, the woolly rhinoceros instead of the forest species, the mammoth and bison of the steppes instead of aurochs. The change was, of course, very gradual and was evidently taken in their stride, for there was still time for art-work, which reached its apogee in polychrome cave-paintings at this time. South of the Pyrenees, where warmer conditions continued to provide a more temperate fauna, even the straight-tusked elephant was occasionally pictured in the caves of Cantabria in northern Spain.

At some such point in time, and while the glacial sea-level was low enough to expose the Behring land-bridge, the eastern Asiatic representatives of Upper Palaeolithic man must have made the first crossing into Alaska. Two factors in this momentous peopling of a whole new Continent stand out. It could not have happened until man, the hunter, had become cold-adapted himself and had become accustomed to hunting the animals of high latitudes. Secondly, it could never have happened at all had there been no ice-free land on the American side during the Last Glaciation, where he could maintain himself until the eventual retreat of the mountain glaciers opened up the way for him to move further in. There was one horn of ice along the shore of the Arctic Ocean and another stretching from the high mountains of north-western Canada through the Alaskan Range and out into the Alaska Peninsula. These enclosed the ice-free Yukon Basin as between the tines of a pitchfork, open to the west, their junction to the east, in the Klondike region, at first blocking further advance. When, now, the ice began to shrink, the rising sea-level would have blocked their retreat in step with the gradual opening of gateways to the mainland New World ahead. They could go forward. They could not go back.

This is not to suggest that the movement of men into the New World was due to purposive exploration. In all probability they advanced in short steps, over many years, from island to island and ridge to ridge, in pursuit of the animal prey – seals, reindeer, whatever it was – on which they lived. Untouched, richer and easier hunting-grounds would lure them on until the Yukon valley enclosed them and became their home – as much as any place can be 'home' to a constantly-moving tribe of hunters. Then, following up the river-trails at last, they would come to a pass and, finding it open and ice-free, would descend beyond it, perhaps into the basin of the Mackenzie River, where no further barrier would stand between them and North America east of the Rockies. This course is quite imaginary, if possible. Their actual routes and the archaeological sites remain almost unknown and unmapped on both sides of the Strait, but in some such way, probably at different times and in repeated waves of immigrants, it must have happened.

Back in Europe, the retreat of Last Glaciation 2 brought in the Second Interstadial, a relatively brief and minor improvement in climatic conditions. The Ice Age was waning, but still had several pauses and slight readvances of the glaciers in store. Gradually more lands were opening up to reindeer-hunters at the Magdalenian stage of culture in north-west Europe. Among the latest of such groups of Late Glacial date were the Hamburgians of Stellmoor and Meiendorf in western Germany and the poor, late outposts of Upper Palaeolithic culture in Highland Britain, called Creswellian from their modest relics found in Derbyshire caves. These were people living at the very limits of the then-habitable world and it is small wonder that their equipment was simple and undistinguished, their art almost non-existent and their time brief.

The major natural changes attendant upon the final retreat of the ice into the mountains of Scotland and of Norway brought corresponding changes into the life of Stone-Age man. As conditions improved, first grassland and heath then birch-pine forest and finally deciduous mixed-oak forest developed in zones over all north-west Europe. The animals of tunda and taiga retreated into the far north, followed, perhaps, by some of the tribes who had long been used to exploit them. Those who dropped behind could not contend with woodlands of increasing density, but betook themselves to rivers, lake-sides, open heaths, beaches and moors, where there remained at least small areas of relatively open country. They adapted their hunting and gathering to smaller and more varied game and doubtless took to an increasing extent to vegetable items of diet which were now plentiful – fruits, berries, nuts and roots in their seasons. These were the Early Postglacial Mesolithic peoples. They had the bow and arrows armed with a variety of specialized points, flaked stone axes for wood-cutting; they hunted the forest red deer, roe and elk, fished and fowled by river and mere and, in hard times and poor seasons, picked up a living by the seashore, collecting shellfish and whatever Nature afforded that men could find to eat. Through the Boreal and the great part of Atlantic postglacial times this was the common way of life throughout most of Europe.

The first economic Revolution had long been prepared in the Near East, where men had first learned to till the ground and sow grain rather than to rely on the harvest of Nature's seasonal seeding grasses. Alongside this activity, and all the changes it necessitated in human habits, grew up another – the herding of beasts, at first captured and tamed from the wild state and eventually bred in captivity to ensure a supply of meat at all seasons for their keepers.

Men with these skills and traditions migrated westwards through Europe, reaching the Atlantic seaboard about 3,000 years B.C. They cleared forest for crops, moving on when the land became exhausted, settling awhile in each new place, taking up more space all the time, chasing away the game and driving the primitive native hunters and gatherers ever further north and west into remoter and naturally less productive areas.

With relatively settled farmers and stockbreeders, the last vestiges of Ice-Age man were displaced and disappeared or were, by degrees, absorbed into the new economy and the new race of inhabitants.

CHAPTER 8

Chronology and dating

The geologist and archaeologist are both closely concerned with the time-dimension in their studies of stratified deposits. The stratigraphy, once worked out, establishes the *order* in which geological and human events took place – the *relative chronology* – and, by means of correlations between strata the time-relations can be determined from deposits over a relatively large area, if the stratigraphical record preserved is sufficiently complete.

Until quite recently (the last half-century) there was, on the contrary, hardly any means available to them, apart from written records, for assigning dates in years to the events of the more distant past, whether natural or due to human agency, which they were studying. It was well enough to know that Event A happened before Event B and that this, in turn, was followed by Event C, but it was obviously just as important further to discover with some exactitude how long ago some of these things occurred and how long were the time-intervals separating them – *absolute chronology*.

Earlier estimates of geological time – not to mention those for the Pleistocene, the youngest and shortest of the Periods – were far too short, and often no more than un-founded guesses. The tendency ever since the development of reliable methods of dating has been to show longer and yet longer dates, far beyond the range of the wildest guesses of the past.

The first serious attempts to date geological time were based on extrapolations from rates of processes which can be observed at the present day. Lord Kelvin (1883) concluded that the probable age of the Earth was about 100 million years, with 400 million and 20 million as possible upper and lower extremes. His calculations were based on measured average rates of loss of internal heat from the globe, extending this back in time to an Earth just solidifying from a hypothetical molten state. He did not, of course, take into account heat continuously being evolved anew in the body of the Earth by radioactive decay, for the phenomenon of radioactivity was not discovered by Marie Curie until 1898. Lord Kelvin's upper limit of 400 million years is to be compared with 4·5 *thousand* million years, the most recent estimate for the age of the oldest known rocks given by Holmes (1965).

For the Pleistocene Ice Ages, Penck & Brückner's (1909) estimates were based on observed rates of sedimentation in lake-deltas, rates of interglacial downcutting of river valleys and the depths to which chemical weathering has progressed on glacial moraines of different ages. Taking the Postglacial as having lasted for 20,000 years so far, they estimated the Last (Riss-Würm) and the First (Günz-Mindel) Inter-glacials at about 60,000 years each and the Great (Mindel-Riss) Interglacial as 240,000 years long. They put the whole length of the Pleistocene (defined as the period of glaciations) at 600,000 years.

It will be seen from what follows that this was a surprisingly good result, in view of the comparative roughness of the purely geological methods available to them.

ASTRONOMICAL THEORY

In 1930, Milankovitch, a Jugoslav astronomer, published curves of the varying intensity of summer radiation received at the Earth's surface in high latitudes during the past 600,000 years. These were based on laborious calculations of the resultant effect of three slight cyclic variations (perturbations) in the Earth's motions, all with different long periods, caused by the remote gravitational effects of the Sun and of the other planets on the slightly irregular figure of the Earth.

The main perturbations are in:

(1) The obliquity of the ecliptic, a variation between extremes of about 21°30′ and 24°30′ in the tilt of the Earth's axis with reference to the plane of its orbit (the ecliptic), having a period of 40,000 years. Its effect is to give more marked contrasts between the seasons at a point on the Earth's surface when the obliquity is greater, but at the same time to lessen the intensity of climatic zonation with latitude.

(2) The eccentricity of the orbit. The Earth's orbit is an ellipse, with the Sun at one focus. The difference in distance from the Sun (and so in the radiation received) between aphelion (the further) and perihelion (the nearer point) every year is greatest when the eccentricity is pronounced, less when the shape of the orbit is more nearly circular. At present, though, from this cause, the southern hemisphere

gets more intense radiation in its summer season than does the north, the summer fraction of the orbit is 7½ days shorter than the winter portion, so that, in the north, though summer at present occurs in aphelion, the northern winter is by that much shorter every year than the summer. The period of change of the eccentricity is 92,000 years.

(3) The precession of the equinoxes. Owing to the oblate figure of the Earth (flattened at the poles, bulging at the equator), the Sun's gravitational attraction of the equatorial belt tends to pull the Earth's axis of rotation towards the vertical in relation to the plane of the orbit. This external force applied to a rotating body results in a slow conical movement (precession) of the axis of rotation, as it does in the case of a spinning top. The effective period of the precession is 21,000 years, which means that the line joining the equinoxes on the Earth's orbit will make one complete revolution of the orbit in this time. Thus, at once extreme (the present situation), the Earth passes through perihelion in the northern mid-winter; with a quarter turn of the precession, it would be at the Spring Equinox; at its half-way point, at northern midsummer; with three-quarters, at the Autumn equinox and, completing the revolution, return, after 21,000 years to perihelion in midwinter.

These three concurrent variations, of very different periods, at one time reinforce each other, at another stand in opposition in their effect on the radiation received at a given point on the Earth's surface. The intensity in the radiation received at that place will therefore vary considerably in time. Calculated for Lat. 65°N (the latitude of the centres of Pleistocene ice-accumulation in Scandinavia and North America) Milankovitch produced a curve for the last 600,000 years of the resultant changes in summer radiation. This dated curve shows several strong minima occurring at irregular intervals, of which the number and spacing correspond surprisingly well with the geological record of Pleistocene glaciations and interglacials shown by the work of Penck and Brückner. It is considered by many that the correspondences are too close and too numerous to be fortuitous and that there must be some causal relationship between results obtained by such entirely different and independent methods. Zeuner, the latest supporter of the Astronomical Theory, insisted that the radiation-variations could not explain the fundamental causes of the Pleistocene glaciations, but might well have been large enough to control the multiple advances of the ice-sheets and their retreat to give the intervening interglacials and inter-stadials. Granted so much, the dates of the repeated radiation-minima could be used, within limits, to date the various maxima of glaciation.

The effect of a period of low summer radiation, and the accompanying high winter radiation, in high latitudes, postulated by the Milankovitch Theory, is to lessen summer melting of ice and increase precipitation of snow in the milder winter, so favouring ice-accumulation and eventual glacial advance. The absolute effects, in terms of temperature and lowering of the snowline, concern climatologists. From the purely geological and chronological standpoints, it seems justifiable to correlate radiation-minima with glacial advances and so transfer the built-in dates of the minima to the latter (Fig. 41).

There is evidently one major source of error in such correlations. Even if we concede that the incidence of a minimum of radiation brought on a renewed advance of the ice-sheets, there must be a considerable time-lag between the cause and its effect – a certain delay, hard to estimate, between a radiation-minimum and the ensuing glacial maximum, probably of the order of one or more tens of thousands of years. Zeuner called this the 'retardation', and, because he had no means of evaluating it, left it out of account in transferring the radiation-dates to the geological phenomena, while still always bearing in mind that the dates must, from this cause, always be somewhat too high.

For twenty years or so the Milankovitch radiation-curves provided the only credible method of absolute dating for the Pleistocene glaciations. Since the 1950's, various radiometric and other methods have been developed, which may be applied directly to samples of geological materials – in particular to igneous rocks by the Potassium/Argon method and to organic matter by the method of Radiocarbon.

Large numbers of Radiocarbon results have by now been obtained, for deposits of the Last Glaciation at the further limit and well up into historic times at the nearer. These check each other so well (even if some of them, individually, remain suspect on particular grounds) that there can be no reasonable doubt that they give fairly accurate dates. These are invariably much lower than those suggested by the Milankovitch radiation-curves. One cause for this discrepancy certainly lies in the retardation, which, for dates within the last 40,000 years, may well represent 25 per cent. or more of the value indicated. Beyond the range of Radiocarbon, however, for dates between 40,000 and 300,000 years B.P., above which figure Potassium/ Argon may give direct dating checks, we still have no more reliable absolute dating method than that of Milankovitch. For such higher dates, the possible error of 20,000 to 30,000 years in some cases, due to ignoring the retardation, becomes less important and it is notable that, in the earlier part of the 600,000 years covered by the original radiation-curve, such Potassium/Argon dates as we now have do correspond remarkably well with those obtained from the curve. This fact increases our confidence in the radiation-curve for the part of the time-scale for which there is no independent check available.

Because of the admitted discrepancies between the radiation-dates and the results obtained by radiocarbon, it has become the fashion, especially in the United States, to abandon the Milankovitch theory as entirely valueless. This is evidently premature and unwise, since there is nothing which can replace it so far in the 250,000-year-long gap in the Middle Pleistocene which falls between the effective ranges of Potassium/Argon and Radiocarbon.

Milankovitch dates for the last million years (for the remoter 400,000 years certainly less accurate than for the latter 600,000) subdivide the Ice Ages as follows:

Major minima at: (1,000's of years B.P.)	*Correlatable with glaciation(s):*
929	
835	Three or more Donau glaciations of Eberl?
762	
720	
590	Günz I
550	Günz II
476	Mindel I
435	Mindel II
230	Riss I
187	Riss II
115	Würm I
72	Würm II
25	Würm III

RADIOACTIVITY METHODS

There are now several methods available of absolute geological dating, based on rates of radioactive decay of particular isotopes of elements, widespread in rocks and minerals, of which the half-lives are known with some accuracy. Two or three of these, only, are of interest for dating the Ice Ages, since the very long half-lives of the others fit them mainly for estimating the ages of materials coming from much earlier parts of the geological column.

Radiocarbon. The best known and most widely used radiometric dating method is that of Radiocarbon (C^{14}, Carbon Fourteen, as it is usually designated by geologists and archaeologists, ^{14}C as it is more properly written in nuclear shorthand).

The theory is as follows. Cosmic radiation reaching the Earth's upper atmosphere includes neutrons. If such a neutron (atomic weight 1, no charge) collides with a nitrogen atom (N^{14}), it is absorbed by the nucleus, forming an unstable body which immediately emits a proton (at. wt. 1, positive charge). The product, though of the same atomic weight as before (14), has lost a nuclear charge with the emitted proton and now has, therefore, a net negative charge, a supernumerary electron. It is thus no longer nitrogen, but has been transmuted into a radioactive heavy isotope of carbon, C^{14}. (Normal, inactive, carbon has at. wt. 12). Though having a different atomic weight, the radioactive isotope behaves chemically exactly like normal carbon, is oxidized by the oxygen of the atmosphere to carbon dioxide, CO_2, and forms part of the Earth's reservoir of this compound.

Radiocarbon decays, by loss of the supernumerary electron, with a half-life period of 5,568 years. Net loss by decay is, in the long term, exactly balanced by addition of radiocarbon newly generated by the neutron flux from space, so that a long-standing equilibrium has been established. In any quantity of atmospheric carbon dioxide, therefore, there is a constant small proportion of radiocarbon and it is assumed that this proportion has not varied within the spans of time we wish to measure. It has recently been shown that this assumption is not, in fact, strictly true, but the variations indicated have been small and do not invalidate the method.

Carbon dioxide from the atmosphere dissolves in rain-water and is mainly stored in the oceans, whence, by one route or another, it becomes available for photosynthesis by green plants. The small proportion of $C^{14}O_2$ is absorbed along with the rest and enters into all the carbon compounds of living organisms, animals deriving it from plants and from each other.

While an organism is alive, it is in constant carbon-exchange with its environment, so that its tissues contain C^{14} in the environmental concentration. When it dies, on the other hand, this exchange ceases and, if any of its carbon compounds are preserved, the radiocarbon in them slowly decays and cannot any more be replaced. If, then, the residual radiocarbon in an ancient organic tissue can be measured, the half-life being known, we can calculate the length of time elapsed since the death of the plant or animal which the sample represents.

Since the amount of C^{14} present in any sample is halved after 5,568 years and quartered after another 5,568, the residual radioactivity after the lapse of (say) 5 half-lives (27,840 years) is reduced to but 1/32nd (2^{-5}) of its original intensity, and becomes only barely detectable. The practical dating-range of radiocarbon is thus confined to the last 30,000 years or so. This takes us back well into the Last Glaciation, but no further – perhaps only 1/100 of the time comprised in the Pleistocene as a whole. The method is most useful over the time-range 20,000 to 2,000 years B.P. and so has been of enormous value for dating Late Glacial and Postglacial organic materials. It will not serve for the greater part of the Ice Ages proper.

Potassium/Argon (K^{40}/A^{40}). A radioactive isotope of Potassium (K^{40}) enters in small proportion into the constitution of potassium-containing rock-forming minerals, along with the much larger amounts of K^{39}, the common, inactive isotope. In the more acid types of igneous rocks, such as granites and rhyolites, potassium-minerals, among them sanidine and biotite, lend themselves to radioactive dating.

K^{40} decays by two distinct routes into Argon (A^{40}), an inert gas, and Calcium (Ca^{40}), with the very long half-life of 1,300 million ($1 \cdot 3 \times 10^9$) years.

When a rock-material has been melted by sub-crustal heat, any free argon which it previously contained will have been driven off, so that the accumulation of argon by radioactive decay of any radioactive potassium present will begin anew from the moment of crystallization of its potassium-minerals.

Obviously, the long half-life of K^{40} best adapts it to the dating of extremely ancient rocks, but the technique of measuring the radiogenic argon has recently been refined (Evernden & Curtis (1961); Gentner & Lippolt (1961)) to the point at which the very small amounts of A^{40} present in rocks of Pleistocene age can be measured with some degree of accuracy, up to dates about 0·25 million years ago. On the Milankovitch radiation-chronology, this puts us at a time somewhere in the Great (Mindel-Riss) Interglacial. For dates between 250,000 years ago and the practical limit of radiocarbon at about 30,000 years ago, a period of some 220,000 years, including most of the Middle and Upper Pleistocene, we have thus, at present, no method of absolute dating save the Milankovitch chronology. It is of some interest, therefore, to compare the results obtained by potassium/argon dating of igneous rocks with those of the radiation-dates, in the area of their overlap in the Middle Pleistocene. Some of them appear in the Diagram, Fig. 39.

It appears that, where they are available, the K/A-dates give results of a satisfactory degree of correspondence with the radiation-chronology. The theoretical error by retardation, of a few tens of thousands of years at most, applicable to the

K

radiation-dates represents, in this area at least, only a negligible fraction of the periods of time involved, a fraction lying, probably, well within the probable margin of error of the K/A determinations.

The practical limitation of the potassium/argon dating method which is most serious is that it is applicable only to a small number of suitable minerals in regions where the stratigraphy includes igneous rocks, whether lavas or pyroclastics. It has, however, proved most valuable for dating in places such as Iceland, Java, East Africa, the Rhineland, Italy and Mexico, where there has been Pleistocene volcanic activity.

Beyond the 1-million-year range of the radiation-curve of Milankovitch, K/A-dates have been obtained for events in the earlier Pleistocene, which are illuminating. That of 1·75 million years B.P. (the mean of several determinations) for Olduvai, Bed I, in an area of the time-scale previously inaccessible to absolute age-determinations of any accuracy, gives an idea of the surprising antiquity of the earliest known Hominids. It now seems, therefore, that the Pleistocene (including the Villafranchian) may well be between 2 and 3 million years long and that the Ice Ages, therefore, occupy, at the outside, only the last one-third of this time.

Protoactinium/Thorium. A recently (Rosholt & Emiliani, 1961) developed radiometric method of dating has been applied to deep-sea sediments. Protoactinium (Pa^{231}) and Thorium (Th^{230}) are both present in the sediments at their deposition. Since these radioactive elements have different half-lives, the ratio between them in a sediment, changing with time, is a measure of its age. The present limit of the method is at about 175,000 B.P. Earlier dates than this for samples from deep-sea cores have been estimated by extrapolation of dated lengths in their upper parts. An assumption is necessarily involved: that the rate of sedimentation has been constant throughout the formation of the column of sediment. This may be a justifiable assumption in selected cases, but in many, because of disturbance or changes in sources of sediment, it is not so, and this is likely to lead to serious errors. The reliable range of the method is, in any case, only a little greater than that of C^{14}, but within this range their results correspond acceptably.

In combination with the study of the changing environmental requirements of Foraminiferal faunas and that of varying surface-water temperatures, gained by o^{18}/o^{16} isotopic analysis of their carbonate, Pa/Th-dating has given (Ericson & Wollin, 1964) a dated sequence of Late Pleistocene and Postglacial climatic events which resembles the terrestrial sequence, and that suggested by Milankovitch, fairly closely. For deeper parts of ocean cores the correct correlations of the temperature-changes indicated with the terrestrial sequence are still a matter of some controversy. Emiliani (1956), for example, has favoured a short time-scale of only some 300,000 years to include all four main Pleistocene glaciations, whereas Zeuner (1959) maintained that Emiliani's curve corresponded only with the latter half of the Milankovitch curve and reproduced also the various minor oscillations of the last two glaciations, Riss and Würm, only (Fig. 41). More work and better dating-methods will finally resolve such gross differences of opinion.

CHRONOLOGY OF SEA-LEVEL CHANGES

Zeuner's straight-line graph of steadily-falling Pleistocene sea-level, drawn on the Milankovitch radiation time-scale, with the interglacial eustatic sea-level maxima marked at their appropriate heights, was the first attempt at absolute dating of the Interglacials, as opposed to the glaciations. It has been pointed out, by Dury (1959) and Holmes (1965) among others, that the slope of Zeuner's graph may be over-estimated. Any decrease in the slope of the line lengthens the time-scale involved. However justifiable as an experimental hypothesis, this at once divorces the graph from the Milankovitch chronology on which the original slope was based. As emphasized here, this is the very premature discarding of a chronological crutch, for which there is, at present, no substitute in the area of the Middle Pleistocene, and one which, moreover, is well supported by the K/A-dates so far available.

K*

CHAPTER 9

Causes of glaciation

Why should the Pleistocene, apparently alone among the geological Periods, have produced an Ice Age?

During the greater part of the last 500 million years, the deposits of which contain fossil remains of plants and animals to guide us in judging this question, it appears that world climatic conditions, even in quite high latitudes, have been equable and warm, indeed, subtropical, in character. Had there, in fact, been, at any time, very widespread and long-continued departures in general ambient temperatures from the range at present tolerable to invertebrate life – something well below the boiling-point of water and at least a little above its freezing-point – it is certain that life on this planet would long since have become extinct. The very fact that life did become established – certainly a lengthy process – and has evidently maintained itself continuously, in ever-increasing diversity and specialization, up to our times, is an argument for the relative stability of world temperatures, since the Precambrian, within the rather narrow range required.

Though, at different times in the palaeontological record there have undoubtedly been periods of great environmental stress for living creatures, during which widespread extinctions of species, and even of larger zoological groups, took place, these seem never to have been so catastrophic as to have involved the whole globe simultaneously, or, indeed, to have constituted more than a temporary check to the increasing variety and adaptation of the surviving forms of life.

If not cataclysmic, there have nevertheless been widespread – perhaps, locally, even violent – changes in the course of the geological history of the Earth. Long spells, lasting for hundreds of millions of years, during which shallow, warm oceans covered most of the face of the globe, have several times been interrupted by lengthy phases of crustal movement, mountain-building (orogenesis) and the emergence from the oceans of wide continents, whereafter relative stability was restored for a further long period.

Once lifted into extensive continents and mountains, the relief of the land and the diversion by it of winds and ocean currents would have altered the regular circulation of heat on the surface of a more or less uniformly oceanic globe and have introduced considerable variety of climate to the new land-masses. High ground would be colder, by reason of its mere elevation and the interior of continents show striking seasonal extremes of temperature.

Prevailing onshore winds in middle latitudes would give high moisture to their western margins, while, in the sub-tropics, offshore winds would result in desert conditions under generally clear skies and hot sun. We live today in just such a stage of relative continentality and high relief, following on the Alpine orogenesis which took place in the mid-Tertiary. The whole of the Pleistocene, of course, shared these physiographic conditions.

After an orogenesis, the land is exposed for long ages to the agencies of weathering and denudation – frost and thaw, wind and rain, streams and rivers cutting into the land-surface, ocean waves eating into coasts and glaciers grinding down the surface rocks inland. Peaks and ranges are first reduced, in time, to low hills, then to slight eminences and finally, in theory at least and given unlimited time, to an almost level 'peneplain'. All the rock-waste from the land is ultimately carried into the sea, to form new sediments in the subsiding ocean basins known as 'geosynclines' and, if the sinking floors of these do not keep pace with the access of sediments, the depressions are filled, water displaced and the seas rise commensurately to flood the lower coast-lands of the continents (transgression). This cycle takes some hundreds of millions of years for its completion, if it is not, meanwhile, interrupted by a fresh upheaval of continent-building scale.

The accumulation of snow and ice anywhere on the Earth, even right at the Poles, is inconceivable during the prevalence of oceanic world conditions. Ready access of convectional heat from lower latitudes would prevent the local maintenance of low enough temperatures. Glaciation obviously requires considerable land-areas in relatively high latitudes (say, more than 55°N. or S.), and preferably high mountains, to initiate snow-accumulation and to provide a gradient for the outflow of the eventual glaciers. Mountains, in these latitudes, also assure sufficiently high precipitation to maintain the focal supply of snow. Low temperatures alone, without abundant moisture, are evidently incapable of initiating, or maintaining, a glaciation. An ice-sheet begins to spread only when each year's access of new snow exceeds the loss of water by summer melting at the ice-edge. Colder winters alone do not suffice:

there must be annual heavy snowfall, combined with cooler summers, slowing down the melting, to reduce wastage.

It is hard to see how an Ice Age could begin, even in the mountains in high latitudes, without some preliminary lowering of present ambient temperatures on the Earth's surface. In the case of the Pleistocene glaciations, as at the present day, it has been estimated that average annual temperature for the world as a whole was in the neighbourhood of 10°C, or 50°F. During the greater part of geological time, it was, on the fossil evidence, perhaps 21°C, or 70°F. Other similar temperature-estimates have most recently been made by means of oxygen-isotope analysis (p. 104), using the tests of oceanic Foraminifera from deposits of the Upper Cretaceous, dating from the end of the last long oceanic phase before the onset of the Alpine orogenesis, in both California and central Europe. They indicate ocean surface-temperatures between 20° and 25°C (68° to 77°F).

Throughout the Tertiary, temperatures indicated by comparable samples fell, on an increasingly steep curve, to near the modern figure. Even ocean bottom-temperatures, as shown by samples from cores taken in the north Pacific sea-floor, exhibit a similar drop, from about 14°C to near freezing, 2°C (57° to 36°F) at the close of the Pliocene. The continued cooling of the oceans, in advance of the Pleistocene glaciations, was, then, a reality, though its cause remains unknown.

Extra-terrestrial influences, such as long-term variations in the output of solar radiation, have been invoked as possible causes of such secular cooling. Short-term fluctuations, associated with the 10·3-year sunspot cycle, are indeed known, but variations with periods of tens, nay hundreds, of millions of years, remain only speculative possibilities, being beyond our observation.

There are other possible influencing factors, this time terrestrial and affecting only the amount of radiation *received* at the Earth's surface, the Sun's output being assumed to remain constant. Atmospheric carbon dioxide (CO_2) readily admits the shorter wavelengths of solar radiation, in the visible parts of the spectrum, but is an effective absorbent filter for the relatively long wavelengths of infra-red radiation emitted to space by the sun-heated Earth. An important increase in the CO_2-content in the past might have increased the heat retained by this 'greenhouse-effect'; a comparable decrease permit more rapid loss of heat by radiation to space. Nothing is certainly known about variations of CO_2-concentration in the past, though there are indications, in systematic variations in Radiocarbon dates for the last few thousands of years that the Earth's CO_2-reservoir – mainly the oceans – may have varied slightly. It is further supposed that it was an abundance of free CO_2 in the atmosphere, combined with high temperatures and plentiful moisture, that evoked the first great explosion of terrestrial plant-life in the Carboniferous.

In any case, the effect of warming the oceans and the atmosphere would be to increase evaporation of surface-water, which, in turn would give greater cloudiness. Clouds reflect back into space about 70 per cent. of the radiation falling on their upper surfaces, so that increase both of CO_2 and water-vapour might be expected shortly to cancel each other's effects and to result in an equilibrium-state no different from the former.

The presence of much dust, in and outside the Earth's atmosphere, should, theoretically, have some climatic effect, either by reflecting some radiation or by providing condensation-nuclei to increase precipitation in the lower atmosphere. The paroxysmal volcanic eruption of Krakatao, in 1883, filled the Earth's atmosphere with dust for months afterwards, without producing any observable systematic lowering of terrestrial temperatures. Eruptions of that intensity must be rare and their effects ephemeral. Long-continued vulcanism on a comparable scale *may* have occurred in past geological times, but, again, this explanation remains speculative. The belt of fine-grained meteoric débris, swept up from space by the Earth and recently detected by satellites, may, similarly, have some influence, but its possible variations in concentration with time can at present only be guessed at. Hardly any of our data bearing on this subject are quantitative, so that we cannot yet decide whether any of the theoretically possible influences is large enough to have produced the observed effects.

It is not, in fact, true that the Pleistocene Ice-Ages constituted a unique series of events, though, being the most recent of their kind, the geological evidences which they have left are widespread, well preserved and prominently exposed. The last pre-Pleistocene Ice Age was in the Permo-Carboniferous, 350 to 270 million years before present. Holmes (1965) lists ten or more Precambrian glacial periods, at irregular intervals between 2,600 and 600 million years ago, of which traces have been reported from widely scattered parts of both the ancient northern (Laurasia) and southern (Gondwanaland) super-continental land-masses.

South and central Africa seem to have been the most frequently-glaciated areas of all. We may content ourselves, here, with looking at the Permo-Carboniferous example, but it is worth noting, in passing, that taking all the rest into account also, there seems to be no regular periodicity, either in ice-ages or in the incidence of orogenic periods, in the dated geological column. Though the Permo-Carboniferous glaciations, like those of the Pleistocene, follow an orogenesis (the Hercynian) at a geologically fairly short interval, this is not the case with most of the Precambrian glaciations, and is certainly exceptional, though the fact may have been effective in reinforcing an already-existent tendency for an Ice Age to develop. In other words, glaciation is not dependent, as has been surmised by some workers, on a preceding orogenesis.

Some other factor than mere pronounced relief of the land is obviously at work, and this seems to be a matter of latitude, combined with the particular distribution of land and water resulting from earth-movements, which favours glaciation in high latitudes if it happens to restrict ocean and atmospheric circulation.

The known evidence for Permo-Carboniferous glaciation all comes from lands which once formed part of the ancient Gondwanaland super-continent – from Argentina, south and central Africa, India and southern Australia. The greater part of the large area affected lies close to, or within, the present-day Tropics. It is a striking fact, in contrast, that the contemporary deposits in the northern continents, the Coal Measures of North America, Europe and Asia (save Arabia and India) indicate tropical-forest conditions in what are now temperate, or even arctic, latitudes, and the succeeding Permian red sandstones (named from a province in northern European Russia) represent hot deserts! No explanation of these hard facts can be imagined, in view of the present distribution of those lands in relation to the existing Equator and Poles and the resultant climatic zones of today. It seems inescapable, therefore, that either the Poles or the continents must have shifted since the Permo-Carboniferous.

The theories of Polar Wandering and of Continental Drift are both of a certain seniority, but have only quite recently (1956) become 'respectable' among geologists. Modern geophysical opinion as to the properties of the Earth's 'mantle' admits that it would permit rheid (viscous) flow, and so make possible movement of the rigid crust upon it. The results of palaeomagnetic surveys (p. 104) now indicate former positions of the Poles in relation to the present continents very different from those of today.

Strictly speaking, the Polar Wandering theory considers that the outer shell of the Earth, ocean-floor crust as well as the continental, has moved as a whole over the mantle, thus displacing its components in relation to the Poles and the axis of rotation, though they remain fixed in the shell and immovable in relation to each other. Continental Drift, on the other hand, maintains that it is the continents that have moved among themselves. Both kinds of movement are theoretically possible and the fact may well be that there have been contributions from both. If, however, Polar Wandering alone had been responsible, the results of palaeomagnetic surveys in all continents should agree as to the positions of the magnetic Poles at the same period (e.g. the Permian) in past geological time. This is not the case. Each continent has given a fairly well-grouped and convincing series of readings representing the past positions of the magnetic Poles at a given period *from its point of view*. The positions indicated by plots from different continents are, however, *all different* – strong evidence that there has since been some relative movement between the continents, not merely wandering over the mantle of the entire shell.

Francis Bacon, in 1620, and many other authors before the present century, have recognized that the outline of west Africa would closely fit the eastern coast of South America, the bulge of Africa at the Tropic of Cancer filling the Caribbean Sea. Alfred Wegener (1915) first published a reasoned and documented presentation of the Theory of Continental Drift, in which he reassembled the present continents round Africa as a centre and positioned a single Lower Palaeozoic super-continent, which he called 'Pangaea' (All-Earth), well to the southward of Africa's present position, so that the Cape lay a little to the west of its present meridian, the South Pole being located in the Indian Ocean off Durban. This arrangement, with the other southern continents and peninsular India in addition, closely grouped round the Pole, explained not only the present distribution of Carboniferous glaciations, but at the same time brought the main coal-bearing deposits of the northern continents into the contemporary Equatorial zone, shifted the present-day north-African desert belt from the Tropic of Cancer to that of Capricorn and created a new north-tropical zone extending from California, across Baffin's Bay and, through Greenland, to Spitsbergen. The contemporary North Pole would then have lain in the north-western Pacific.

This conformation, or something like it – for later authors have modified Wegener's arrangement to suit themselves, without altering his general principle – seems to have persisted from Precambrian times into the Middle Mesozoic, the drift from the Poles and present separation of the continents having largely taken place since then. If we take the distance between easternmost Brazil and Nigeria, for instance, as being 3,000 miles in round figures, the time available for drifting is somewhere about 200 million years and the average rate of steady drift a quite modest figure of 1 mile in 6,000 years, more or less – something like 1 inch annually!

ICE AGES: THEIR NATURE AND EFFECTS

By now, very large numbers of factual data, from structural geology, petrology, oceanography, geophysics, besides the long-standing problem-facts of plant- and animal-distribution, which have for years plagued biogeographers and palaeontologists, all stand in support of Continental Drift as a workable – and working – hypothesis.

The point of all this, from the present angle of interest in the causes of Ice Ages, is dual. First, accumulation of ice is shown to have taken place, at many points in geological time, even on comparatively low-lying, but very extensive land-masses, provided that they are located mainly in suitably high latitudes – in this case, near and surrounding the South Pole. Secondly, since those glaciated lands have since drifted far apart, Continental Drift may have been principally responsible for producing the set of geographical circumstances which made possible the initiation of the Pleistocene Ice Age of the northern hemisphere.

In addition, the evidence of radiometric dates for the Permo-Carboniferous glaciations suggests that they began in the west (South America), traversed south Africa and India and ended in Australia in the Permian, indicating that the Gondwanaland continent was, during the long period of multiple ice-advances and ice-retreats, actually drifting across the Pole.

One serious past objection to differentially-drifting continents was the inability of proponents of the theory to point to possible forces tangential in direction to the Earth's surface competent to move such enormous masses against the frictional resistance of a very viscous mantle supporting them. A combination of the resources of modern oceanography (including sensitive gravity-surveying, detailed echo-sounding profiles and so on) and of palaeomagnetic surveying aided by radiometric dating, has revealed the structure and properties of the well-known mid-ocean ridges, in particular of the North Atlantic Ridge. This is a wide, more or less N.-S.-trending irregular linear elevation of the ocean floor, with a deep rift running along its centre. Rock-samples dredged from the rift prove to be geologically young, but their age increases regularly both to east and west with increasing distance from the centre. It appears, therefore, that the rift is a real crack in the ocean-bed crust with continually-emerging new rock-materials rising in its centre and spreading laterally, thus gradually widening the Atlantic basin and slowly forcing the Old and New Worlds further apart. It appears to be due to the upwelling of a very slow, viscous convection-current in the underlying deep layers of the crust or upper mantle, and not only provides a quite sufficient tangential force to account for the drift, but, on the magnetic and dating evidence, must have, in fact, had that result. In other places, notably along the west coast of North America, there is a corresponding sinking of ocean-bed crustal material beneath the margin of the continent. There is now, in slow continental drift, a perfectly satisfactory explanation for the Permo-Carboniferous glaciation of regions nowadays in the Tropics – they were at that time circumpolar!

During the Pleistocene Ice Age, however, the distribution of land and water did not differ essentially from that of the present day. Continental Drift at a constant rate takes tens or hundreds of millions of years to result in important displacements. The Atlantic Rift would, thus, have widened by only 30 miles during the whole of the 2-million-year-long Pleistocene. We cannot, therefore, use Continental Drift directly to explain the Pleistocene glaciations, though, as pointed out above, it may, incidentally, have been responsible for the circumpolar arrangement of the continents. We must, then, use a close approximation to present-day geography as the basis for our consideration of the possible causes of the latest Ice Age.

Surviving ice-sheets are found today only in Greenland and Iceland, in the north, and in Antarctica. The latter is a land-mass of continental dimensions and high relief centred on the South Pole and completely surrounded by oceans – an ideal situation for ice-accumulation. The North Pole, on the other hand, is situated somewhat eccentrically in an almost land-locked ocean of a size comparable with Antarctica. In the course of the drift of the American continent since the Permian, Alaska seems to have passed right over, and beyond, the North Pole, in a south-westerly direction. In seeking for a purely terrestrial cause for the initiation of Pleistocene glaciation in the north, this arrangement is crucial.

Ewing & Donn (1956) published a theory of glaciation, based on this fact, which has much to recommend it.

The relative isolation of the Arctic Ocean by land-barriers, broken only by shallow straits, limits exchanges of its waters, by convection, with those of lower latitudes. With the North Pole in its former position, in the north-west Pacific, the then polar part of that Ocean could never freeze, because of free heat-exchange through deep water, with the rest of its mass.

It has, further, been suggested (Stokes, 1955, 1957) that a latest Tertiary or earliest Pleistocene uplift of the Panama Isthmus, formerly an open chain of volcanic islands, prevented leakage of warm water into the Pacific from the Caribbean Sea and the Gulf of Mexico, so that a stronger-than-before Gulf Stream and North Atlantic Drift was sent north-eastward. The south-westerly winds prevailing in the north temperate zone would thus be warmed and carry plentiful marine heat and moisture northwards, towards the present Arctic Circle. A certain amount of this circulation of tropical origin may have reached the basin of the Arctic Ocean via the wider gaps between Greenland, Iceland and Norway, keeping that Ocean ice-free, but the moisture-laden winds, striking the cold higher ground of the northern land-masses, would result in heavy snowfall and begin the growth of glaciers.

Snow and ice, covering the land, would increase its 'albedo', the reflectivity for incident radiation. As much as 70 per cent. of available radiant heat would thus be sent back into space instead of being absorbed. This would result in even lower temperatures and reduced melting, even in summer. An ice-sheet, once started, thus tends to provide its own impetus for growth.

Not only so, but the meteorological effect of a large area of land covered with ice is to cool the air-mass over it and this dense, cold air, at some critical size in area, begins to form the centre of a high-pressure system, the 'glacial anticyclone', from which cold, dry winds radiate, absorbing heat and moisture from adjacent regions not yet ice-covered, the 'periglacial zone'.

Warmer, moist air-masses, reaching these chilled margins from lower latitudes in the Atlantic would give up their water-content as heavy snow, thus adding to the accumulation favouring spread of the ice. Both lowered sea-level on the continental shelves, due to the withdrawal of the ocean-water to feed land-ice, and the accumulation of the ice itself, in some places amounting to thousands of feet in thickness, would raise the effective height of the land and increase its distance from the new sea-shore, lowering temperatures and militating against melting and wastage.

The ice-edge, now invading much lower latitudes, would cause a steepening of the temperature-gradient between itself and the adjacent, still relatively warm, oceans, promoting more intense circulation by convection both of air- and water-bodies. The resulting increased storminess, due to contact of air-masses of very different temperatures and humidities, would bring about heavy marginal precipitation, much of it as snow.

These self-reinforcing processes cannot go on increasing indefinitely. By now extending, at its maximum development, far from its centre of accumulation, the ice had little mechanical impetus for further spread, save its own dead weight, so that wastage by summer melting under the warmer sun of lower latitudes would begin to balance advance of the ice-front. The volumes of cold water returned to them would have cooled the neighbouring seas considerably, enabling the Arctic and North Atlantic Oceans to freeze over. By preventing evaporation from a free water-surface, this would cut off the supply of north-bound moisture in great part and, at the same time, by increasing the area of high albedo – a frozen sea reflects far more heat than calm open water – would markedly increase the extent of the glacial anticyclone. This would now tend to divert approaching westerly depressions in the north temperate zone to a more southerly course, even further reducing precipitation over the ice-sheet.

Deprived of its motive force, in supplies of new snow, wastage at its edges would soon surpass the rate of spread of the ice and retreat from the maximum begin, but not for thousands of years more would enough heat accumulate in the oceans, continually fed with meltwater from the thawing land-ice, to reopen the Arctic and make it possible for a renewed glaciation to begin. Even today, though the deglaciation of North America and Scandinavia have been complete for ten thousand years, Greenland still receives enough snow from the Atlantic to maintain its ice-sheet and only the very margins of the Arctic Ocean unfreeze in summer.

The theory of Ewing and Donn provides a credible mechanism for the onset of a northern Ice Age and its eventual decay and retreat, once the necessary geographical conditions had been established. It accounts, also, for the undoubtedly long time-lag in complete reversal after the maximum glaciation, of which we are still the witnesses. We still lack, however, almost all the exact quantitative data that would be required for its proof and there remain numerous problems of detail which cannot be fully solved by their suggested process. Most notable among these is the great complexity of the Pleistocene Ice Age as we now know it, including multiple advances and retreats, some of very long duration and extent, which are not provided for in the grand overall scheme which they present.

For the repeated incidence, within a major Ice Age, such as that of the Pleistocene, of longer or shorter Interglacial and Interstadial periods, we must have recourse to some other mechanism. It would seem that the varying incident summer radiation in higher latitudes, calculated by Milankovitch for the last million years, provides controls of adequate scale to explain in detail the observed advances and retreats of the glaciations. Unwilling as some North-American authors appear to be to accept this, no other equally satisfactory theory has ever been advanced.

Finally, a possible extra-terrestrial cause of glaciation, as well as of other abrupt, violent and perhaps very widespread geological effects, is meteoritic impact.

Most of the solid bodies from space, captured by the Earth's gravitational field and running the gauntlet of entry through its dense atmosphere, are of very modest dimensions, so that few, in fact, survive frictional heating, but burn up before reaching the Earth's surface as meteors. Their solid residues eventually drift down to the surface in the form of almost impalpable dust and fused glassy globules.

The comparatively rare more massive objects, still generally only a few kilograms in weight, survive their passage, at least in part, to be seen and wondered at as 'fireballs', the fall of which is often accompanied by 'supersonic bangs' or actual explosion and fragmentation on impact, with some occasional damage to life and property. These are meteorites, the larger of which, a few hundred kilos in mass, are so uncommon as to have been individually named, placed in museums and studied, as the only undoubted masses of extra-terrestrial matter known to us before the recent American moon-landing. One of these, large enough to evoke widespread interest, alarm and local devastation, fell in Siberia in 1904. Probably 70 per cent. fall unobserved and harmlessly in the oceans.

Unrecorded ancient, and still larger, meteorites have left their unmistakable traces on land in the form of explosion-craters, of which the Barringer Crater in Arizona is probably the best preserved and widely-known example. Unless geologically fairly recent in date, the elevated rim of débris, the entire circular outline and other striking characters may be almost obliterated by denudation and many of the larger of such 'astroblemes' ('star-wounds'), some tens of kilometres in diameter, have remained unrecognized on the ground in remoter areas until their presence was indicated by circular features appearing in aerial photographs. Similar structures, clear enough and long well known to geologists, because close to inhabited places, have not been admitted by them as impact-features, but have been interpreted as 'crypto-volcanic' explosion-craters or cauldron-subsidences. There are now grounds for reconsidering some of these.

Some of the putative meteorite-craters are very large indeed. If they are, in fact, the scars of impacting extra-terrestrial bodies, the probable masses and velocities of these can be estimated and their likely effects on the Earth calculated. Some are probably the marks of random and catastrophic events, often in the remote geological past, due to bodies large enough and energetic enough to have affected the rate of the Earth's spin, to alter the attitude of its axis or, through sudden shift of the crust by slipping over the underlying mantle, to cause displacement of continents relative to the Poles and to each other. It has, thus, been maintained that Continental Drift has been, not continuous, but abrupt and catastrophic.

Sudden and violent as such changes must have been, they could not fail to provoke earthquakes, tidal waves, hurricane-force winds and widespread destruction of contemporary living species. This is the doctrine of Neo-Catastrophism.

Such events cannot simply be dismissed as impossible merely because of their small probability and only occasional apparent incidence in the past – at any rate within historical times.

Impacts of extra-terrestrial bodies large enough to have had such generally catastrophic effects could well so have altered the Earth's attitude and motions, and the distribution of land and water, as to have caused, or at least triggered, the onset of glaciation. As far as concerns the Pleistocene Ice Ages, the onus clearly rests upon the Neo-Catastrophists to demonstrate the location, direction of impact, energy, dating and adequacy of a particular incident to have had the generally observed geological effects. Further, since the fall of meteorites is both rare and evidently random in timing, they cannot seriously be held to account for all the repeated glacial events of the Pleistocene.

A recent book (Gallant, 1964) interestingly pleads for the reality of meteoritic impacts, both in general and with some particular examples, details their consequences and supports the case with a wealth of circumstantial evidence and calculations based on the best available geophysical data and opinion to quantify the evidence. Unlike some other possible extra-terrestrial causes of glaciation here mentioned, the evidence for meteorites is accessible on the ground and should be readily quantifiable to prove their competence to produce the effects claimed.

From the point of view of classical Lyellian geological doctrine, such Neo-Catastrophism is rank heresy and apparently destructive of the geologist's main tenet of Uniform Causes. This is not so. It is plainly apparent that, during most of geological time change has, on the whole, indeed been slow and gentle. The Neo-Catastrophists merely point out that evidence now available shows that, doubtless only at rare intervals, sudden cataclysms on a global scale, due to meteoritic impacts have, in fact, occurred and so must be taken into account. Such impacts are, *ex hypothesi*, random in timing and magnitude, so that, when they *have* happened, they must have introduced disturbances into the regular processes normally alone sufficient to account for geological phenomena. Among other matters of geological interest, therefore, the Pleistocene Ice Ages and their causes cannot any more be exempt from the possibility of random interference by meteoritic impacts.

It is evident that we are, as yet, far from understanding in detail the mechanisms initiating, controlling and concluding Ice Ages, though with the general advance of geological (especially geophysical) knowledge, we think that the problems are not insoluble – indeed that existing theories already go some way towards explaining them.

CHAPTER 10

Conclusion

Forty years ago, although interest in the Pleistocene Ice Ages was already respectably centenarian, most books on general Geology tailed off rather disappointingly in their final chapters and the reader might reasonably gain the impression that, to geologist and palaeontologist alike, the events of the last 1 million years were scarcely worthy of any serious attention. Pleistocene deposits, briefly characterized as glacial drifts, river-gravels, brickearths and peats, were hastily dismissed as 'superficial', with the implication that, besides themselves being really devoid of interest, they had the added disadvantage of concealing the important 'solid' strata beneath. For completeness sake, the author might mention the Günz, Mindel, Riss and Würm glaciations of Penck & Brückner, figure a hand-axe and a Mousterian 'point' and include a borrowed engraving of a (generally bad!) restoration in the flesh of a mammoth or an 'Irish Elk' and so, figuratively heaving a deep sigh of distasteful duty well done, bring his work to an end.

It was the awakening interest in the origins and earliest history of man himself that gave to Pleistocene studies the fillip which started the impetus continuing increasingly to the present day.

However sketchily described here, it cannot but be evident to the reader who has persisted so far that the Pleistocene – and especially the latter part of it which includes the Ice Ages – offers intrinsic interests to followers of many scientific disciplines, whose particular knowledge and skills may profitably be applied to the solution of the problems which it raises. The geologist, the geomorphologist and the student of early man and his works; the climatologist, the glaciologist and specialists in palaeontology (especially of the mammals), pollen analysis, sedimentology, mineralogy and soil-science – all find that their particular interests converge in the Pleistocene, to discover, illustrate and describe in their own terms a great variety of natural environments and local geographies.

ICE AGES: THEIR NATURE AND EFFECTS

These were the backcloth to the earliest stages of the long human drama, which, in the last few millennia before Christ, emerged into the light of History. Even the most dedicated and myopic specialist can hardly fail to appreciate that his results may prove to have an important bearing on some aspect of the earlier history of the human race.

While we are all striving to push back the boundaries of our ignorance in our particular corners of this wide frontier, it is well if we keep before us the fact that it is on the shoulders of our predecessors, or at least in the firm footholds which their work painstakingly cut for us, that we stand to wield our own implements and instruments of research. It is all too easy to look down from the heights recently scaled upon the achievements, diminished only by distance, of the great men who preceded us in the field. In very many cases, time, even with the introduction of new methods to bear on their problems, has shown that their work was sound and their conclusions still valid today. It is often worthwhile to read their original works – not only modern critical summaries of them – and, if these reveal that they were sometimes mistaken, and, on occasion took to new ideas only grudgingly, so probably will we too appear to future workers, looking at our times just as critically from their superior level of knowledge.

On the other hand, there are always those, today, who are ready uncritically to adopt the results of the latest dating method or the re-labelling with outlandish new terms of phenomena long well known. They may too readily abandon, as totally useless, well-authenticated older results and widely-understood established terminology, merely to appear up to date. Fresh results by new methods are always of interest, and cannot be ignored, but, until their reliability has been proved, it is unwise to discard entirely the props we already have. It is like a cripple's accepting a single walking-stick in exchange for a well-tried pair of crutches! So, too, with new names for old things. If they add to our knowledge and understanding of the thing named, well and good: if not, they are inevitably only confusing – we are liable to waste time arguing about labels while being fully in agreement about the phenomena for which they stand.

Pleistocene deposits are relatively slight, being largely terrestrial in origin, extremely local and varied lithologically and often laterally discontinuous, so that firm correlations between exposures are hard to make. In places they are greatly complicated, dissected, contorted and reworked by repeated advances and retreats of ice-sheets introducing erratic materials from elsewhere. In a way quite unlike that of the 'solid' geologist, a worker on Drifts has to rely greatly on topography and land-forms for clues to possible connections between his deposits, on the recognition of erratics, fossil soils and plant-beds to subdivide his stratigraphy and to recognize traces of erosion, disturbance and redeposition of sediments which may indicate the possibility of discontinuities – indeed, often huge disconformities – standing for possibly long stretches of time unrepresented by any deposits in his relative chronological sequence.

The emphasis on European phenomena in this book is unavoidable. Pleistocene studies, after all, mostly began in Europe and, while much work has been done elsewhere, especially in North America, it is in these two continents, above all, that, with highly developed technologies, industrial and civil-engineering operations have opened up rail- and road-cuttings, quarries and excavations for gravel, sand and clay, which have given us the sections in superficial deposits on which much of our knowledge is based. Elsewhere, Nature does often provide magnificent natural sections, especially in the walls of the gullies of temporary watercourses in sub-arid lands, but too often these have hitherto gone unnoticed for lack of trained and inquiring eyes and minds to investigate what they display. The distribution of important Pleistocene sections and sites on the map of the world is often only that of assiduous and dedicated workers of yesterday and today, and there are large areas of which next to nothing is known. With the development of backward lands, increasing political and economic stability, the spread of higher education and the pursuit of knowledge not of direct or immediate commercial exploitability, we shall gradually get to know much more about this latest and most enthralling period of Earth's past.

What of the future? Are we out of the Ice Ages yet? What we call the 'Postglacial' has so far lasted for a mere ten or twelve thousand years. Does not our study of the past enable us to glimpse, at least, the possibilities, if not the probabilities, of what lies ahead?

Even the Last Interglacial (a short one) seems to have endured for something like seven times as long as the Postglacial. Looking at the pattern of floral development for an Interglacial, we have, by now, clearly emerged from the Late Glacial and Boreal stages in north-west Europe into an Early Temperate phase, with dominant mixed-oak forest. If we are to return, eventually, to fully glacial conditions, there should follow, after some further tens of thousands of years at least, Late Temperate and Post-temperate stages, before seriously deteriorating climate accompanies renewed ice-accumulation in the highlands of the north.

If we are in a mere Interstadial, on the other hand, our temperate phase may already be half over, so that mankind in the northern zones can look forward only to increasing cold from now on!

In the former case, it can be expected that, with minor oscillations, deglaciation will, on the whole, continue for a long time to come, with progressive reduction in the area and thickness of the Greenland and Antarctic ice-sheets and, ultimately, unfreezing of the Arctic Ocean. The return of all the meltwater to the oceans will raise world sea-levels, possibly by as much as 100, at least by several tens of metres, in the event of complete disappearance of persistent ice, and this will inevitably be the case if, in fact, we are at the end of the Ice Ages.

The secular fall in sea-level which we have noticed will, no doubt, continue, but, at the most optimistic estimate of its rate (Zeuner's 12 mm. per century) it will have fallen but 1·2 m. in the next ten thousand years and so scarcely at all mitigate the eustatic rise in that time due to continuing deglaciation. As a result, many of the world's low-lying coastlines will be flooded by a renewed marine transgression and many of man's most important civic, commercial, industrial and agricultural works be overwhelmed.

No doubt, by that time, man's way of life will have altered considerably, as it will have to do if he is to compensate for these slow geographical and environmental changes. Archaeologists of the future will, in that case, have to search for the relics of today's civilizations largely under water. A beginning has, after all, already been made in this direction!

Technical advance, even if at a much slower pace than today's, should amply keep up with Nature's deliberate changes. Indeed, if by then man has learned to control weather, and even climate, he may be in time to prevent either total deglaciation and the consequences outlined above, or a threatened recurrence of ice-advance which, apart from itself covering large areas of now habitable land, would, conversely, dry out his ports and turn continental shelves into wide extensions of present lands, while setting all the rivers of the world to work at deepening their channels!

If, on the other hand, man suffers world-famine or wars (or even a meteoritic cataclysm!) he may quickly fall back to a primitive subsistence-level at which environmental control, even of the simplest, ceases to be possible, and so be too late to obviate (or even appreciate) the threats to his future posed by the untrammelled processes of natural law. So the mixed-oak forest will, once more, cover his works in

Europe and the world return to a state something like that in which he first found it.

That state could never be quite the same, for the Pleistocene elephants and rhinoceroses and giant deer have gone for ever, the bison is all but extinct, surviving cattle and horses are of domesticated breeds ill suited for renewed life in the wild and nearly all the larger wild animals are perilously reduced in numbers, even if small breeding-stocks were locally to survive whatever disaster had decimated their worst enemies.

Should man nevertheless survive the prophets' worst prognostications for another ten thousand years, it is likely that, by his technology, he will have almost entirely superseded Nature. Even if left to herself, Nature never repeats herself exactly. We can only guess at general possibilities, but never predict in detail what our world will be like if, or when, it is subjected to another Ice Age.

L

Glossary and Index

ABBEVILLE (N. France), Gravels of R. Somme, containing fossils and human implements, 63, 82, 120

ACHENHEIM (Alsace), Section showing multiple loesses and buried soils formed thereon, 101

ACHEULIAN, Human cultural stage marked by a preponderance of often well-made hand-axes, 72, 82

AFTONIAN, Interglacial period in N. America, between the Nebraskan and Kansan glaciations. Equivalent to the Antepenultimate Interglacial in Europe (ApIgl), 63

AGASSIZ, Louis, Palaeontologist and glaciologist, 13

AGGRADATION, Build-up, by accumulation of sediments, of river flood-plain.
  Climatic a., 30
  Eustatic a., 52

ALBEDO, Reflection back to space of incident solar radiation from the Earth's surface. Dependent on quality of the surface—rock, vegetation, water, ice etc., 39, 157

ALGAE (sing. ALGA), Microscopic unicellular green plants, 45, 47

AMERICA, Glaciation of, Fig. 31
  Migration of early man to, 118, 136

ANIMALS, As environmental indicators (see also FAUNA), Chap. 6, *passim.*

ANTEPENULTIMATE GLACIATION (ApGl), =Mindel/Elster/Kansan, 64
  Interglacial (ApIgl)=Günz-Mindel, Cromerian, 62

Fauna of, 62
Flora of, 63

ARGILE ROUGE (Fr. red clay), Weathering-soil of Last Interglacial on Older Loess, 82

ASTRONOMICAL THEORY of Glaciation, Perturbations of Earth's motions, affecting radiation-receipt in high latitudes, cause multiple advances and retreats of glaciation, 104 *ff.*

ATLANTIC CLIMATE, Oceanic, mild, moist type of climate, without wide seasonal extremes of temperature, 92

AUSTRALOPITHECINES, Hominids, mainly African, of Lower Pleistocene date, belonging to the sub-Family Australopithecinae, 127

BASE-LEVEL, An ideal profile to which a river cuts down or aggrades its bed to attain equilibrium with static conditions, with erosion and sedimentation in perfect balance, 47

BEACHES, Lake-beaches, Fig. 19, 95
  'Raised' b's, 54
  Sea-b's, 35

BERGSCHRUND (Germ.), Term for the head-wall crevasse of a firn (q.v.)-field, 18, Fig. 2

BENCH, Cut in the 'solid' or in pre-existing 'drift', marking the limit of erosion of a river at the end of a period of down-cutting. Terrace-gravels lying on the bench are not, necessarily, of immediately sequent age, 66, 70

166

BISHOP TUFF, Volcanic ash of an Early Pleistocene eruption in U.S.A. Dated by Potassium/Argon, 62, Fig. 28

BLACK ROCK, Brighton. Last-Inter-glacial high beach, 85, Fig. 37

BLUFF, Steep side of a valley, rising above an almost level river flood-plain, 47

BOULDER-CLAY, or Till. An unstrati-fied, unsorted, geological deposit of a glacier or ice-sheet, 12, 21, 95

BOREAL Climate or flora. Literally 'northern' (Lat. Boreas=north wind). Applied, here, to the cold-temperate zone, 66, 112

BOYN HILL TERRACE of R. Thames. Named from locality near Maidenhead. 'Hundred-foot' (33 m.) eustatic terrace of the Penultimate Interglacial, 71

BRANDENBURG Moraines (N. Germany), End-moraines of Weichsel glaciation (?=Last Glaciation II), Figs. 30, 35

BRASENIA PURPUREA, Exotic water-lily, formerly widespread during European Interglacials, 84, 114

BRODEL-SOILS (Germ.)—see also Structure-soils, Permafrost, 31ff.

BUCKLAND, W., Dean of Westminster, geologist, 13

BURIED CHANNELS, Former river-courses, graded to sea-levels lower than at present, now buried by aggradation of the modern flood-plain deposits. Three such, in the lower Thames, correspond to the three low sea-levels of the Last Glaciation, 87, 91

  SOILS, Soils formed by weathering of a former land-surface, now covered by subsequent geological deposits, 49, 88, 101

CALABRIAN Sea-level. Early Pleistocene 200-m. high sea-level and beaches attributable thereto. Named from Calabria, S. Italy, where first recognized, 60

CARBON FOURTEEN (C14), Radio-carbon. A radioactive heavy isotope of common carbon (C12) with a half-life of 5,568 years, used for dating organic materials back to about 30,000 years B.P., 91, 104, 144ff.

CHAMBERLIN, T. C., Geologist, 14

CHANNEL, buried, 35, 87, 91
  English, 64, 92
  glacial-drainage, 24, 25

CHELLIAN, Human cultural stage, typified by rather roughly-made stone hand-axes (Chelles-sur-Marne, N. France), 73, 82

CHILTERN DRIFT, Ancient (?=ApGl₁) boulder-clay with western erratics, capping hill-summits in Hertfordshire, 68

CHOPPER-IMPLEMENTS, formed on pebbles, generally preceding hand-axes in date. Earliest recognizable human artifacts, 128, 130

CHRONOLOGY, Relative and Absolute, 139

CIRQUE (Fr.), Corrie (Scot.) or Cwm (Welsh) – The, often semicircular, high valley-head, like an amphitheatre, once occupied by the firn-field of a valley-glacier, 20

CLACTON CHANNEL (Essex), A former river-channel (? of R. Thames), now close to present sea-level, indicating a down-cutting of nearly 100 ft. (33 m.) between the Lower-Loam and Middle Gravel stages of the lower Thames at Swanscombe, 72

CLACTONIAN, Early human industry and technique, first recognized at Clacton, Essex, consisting of stone flakes detached from a polyhedral core, 70, 77

CLAY-GRADE, Particles in a sediment less than 2 micrometres (0·002 mm.) in diameter, 33

CLIMATIC AGGRADATION, Deposition of an overload of glaci-fluvial sediments by a stream, thus raising its valley-floor under a cold climate, 30

  TERRACE, A terrace, isolated by subsequent down-cutting, under a temperate climate, formed by a stream which originally aggraded the material in a cold period, 47

CONTINENTAL CLIMATE, Typical of regions far from the thermostatic effect of the oceans, in which seasonal extremes of temperature and precipitation are most marked, 124

DRIFT, Theory of. Possible relative movement of continental blocks *inter se* resulting from upwelling of mantle-material at the mid-ocean rifts, 152*ff*.

SHELF, The areas of shallow (less than 200 m.) seas surrounding the continents, divided from the ocean deeps by the continental slope, the true margin of the land-mass, 35, 52, 156

CONTORTED DRIFT, Local name, in N.E. Norfolk, for the moraine of an early East Anglian glaciation (=Elster/Mindel/Lowestoft, ApGl), 66

CORIOLIS FORCE, Force, due to Earth's rotation, tending to alter the course of any body, moving on the Earth's surface, in a clockwise direction (in the northern hemisphere – anticlockwise in the southern), 31

CORRIE (Scots), Cirque (Fr. – q.v.) or Cwm (Welsh), 20

CORTON SANDS, Sands, presumed of interglacial origin, between lower and upper boulder-clays at Corton, Norfolk, 66, Fig. 29

CRAG(S), Local name, in East Anglia, for a series of sandy/gravelly shallow-water marine deposits of Pliocene/Early Pleistocene age: Coralline C., Red C., Norwich C., Weybourne C., 60

CREVASSE (Fr.), Narrow, deep tension-crack or crevice in a glacier or ice-sheet, 20

CROMAGNON MAN, An Upper Palaeolithic race of *Homo sapiens*, first recognized at the Cromagnon Cave, Les Eyzies, Dordogne, 88

CROMER FOREST-BED SERIES, Interglacial river/estuarine deposits in Norfolk (ApIgl), 63, Fig. 29

CROMERIAN INTERGLACIAL, Antepenultimate Interglacial, named from above, 62

Flora of, 114

Fauna of, 120

CROMER TILL, Boulder-clay with Scandinavian erratics of earliest glacial phase (ApGl$_1$) recognized in East Anglia (Inland equivalent is Norwich Brickearth), 66

CWM (Welsh), Cirque or Corrie. High valley-head, like an amphitheatre, once occupied by the firn-field of a glacier, 20

'DEAD' ICE, A mass of ice, detached from, but transiently persisting, before the front of a glacier or ice-sheet, on retreat of the coherent front, 24

DECKENSCHOTTER (Germ.), =cover-gravels. Glacial-outwash plains of early Alpine glaciations, 14

DEEP-SEA CORES, Samples, up to 20 m. in length, of ocean bottom-sediments, obtained by a piston-corer, a hollow tube driven into the sea bed, 61, 102, Fig. 41, 146,

DEFLATION, The process of removal, by wind, of dry, finer sediments from the surface of a sandr, outwash-plain, dune-field or other area of bare incoherent deposits, 28

DENUDATION, General lowering of topography by erosion and transport of comminuted and weathered rock-material, 68, 86, 149, 158

DIATOM(S), Microscopic unicellular plant with a siliceous skeleton (frustule), both of freshwater and marine environments, 45

DIATOMITE, A sediment, floury or silty in consistency, largely composed of diatom-frustules, 45

DIESTIAN SEA, A marine transgression, in the London Basin and Low Countries, of Early Pleistocene age. Its beach-platforms and deposits, at about 600 ft. (200 m.) correspond with the Calabrian stage of the Mediterranean, 60

DISCONFORMITY, A break or gap in a stratigraphical sequence not marked by any angular unconformity (q.v.), 72, 163

DISPLUVIAL CONDITIONS, Rainfall unevenly distributed within the year

L*

MOUSTERIAN, A human cultural stage marked by well-made flint flake-implements, the work of Neanderthal Man, 88

NEANDERTHAL MAN, *Homo sapiens neanderthalensis*, a race of the modern human species which occupied caves in S.W. Europe during the first phase of the Last Glaciation, makers of the Mousterian industries. Extend, as far as is known, at least to Central Asia, 88

NEBRASKAN GLACIATION, in N. America represented by an early boulder-clay, little exposed save where coverings of later drifts have been eroded away. Dated by the Bishop Tuff (q.v.) to 0·87 myr. B.P. Hitherto supposed to correspond with EGl (=Günz) of Europe, but this date puts it well back – even earlier than Donau! – in the Villafranchian.

As, furthermore, the Nebraskan boulder-clay is little less extensive than the maximum glaciations of the Kansan and Illinoian, it is, in this respect also, very unlike the European Günz. It seems possible that several ground-moraines are confounded under the name Nebraskan, 62

NECHELLS, Birmingham, Site of important Great Interglacial (PIgl) deposits with plant and insect remains, 72

NEO-CATASTROPHISM, Modern theory of multiple meteoritic impacts having often sudden and cataclysmic geological results. In conflict with the established doctrine of Uniformitarianism (q.v.), 159

NÉVÉ (Fr.) or FIRN (Germ.), Term for old, re-frozen and consolidated snow, a stage in the formation of glacier-ice, 17

NORWICH BRICKEARTH, An ancient, decalcified, boulder-clay, containing Scandinavian erratics. Probably the inland equivalent of the Cromer Till, evidence of the earliest East Anglian glaciation. (ApGl₁), 66

NUNATAK, Greenland-Eskimo word for an isolated rock-peak emerging above the level of an ice-cap, adopted as a geological technical term, 20, 52

O.D. (ORDNANCE DATUM), The conventional zero-level, representing mean sea-level for the British Isles, from which the Ordnance Survey formerly measured all heights in Britain. It is superseded by 'N.D.' (Newlyn Datum) taken from West Cornwall, which differs from O.D. by +1 ft., 70

OLDER DRIFT, Earlier glacial deposits of S. and E. England, the products of glaciations earlier than the Last Glaciation (i.e. of PGl and ApGl=Riss/Saale and Mindel/Elster), 76

OROGENY, Period of mountain building. The most recent was the Alpine, during the mid-Tertiary, 57, 148, 151-2

OUTCROP, Area of exposure at the surface (i.e. area mappable without excavation or boring) of any geological deposit or formation, 12, 95

OUTWASH, GLACIAL, Deposits of glaci-fluvial meltwater-streams, mainly of gravels, sands and silts, 14, 26*ff.*

OXYGEN-ISOTOPE ANALYSIS, The ratio of O^{18}/O^{16} enables the estimation of ocean-surface temperatures from tests of Foraminifera once living therein, 104, 146, 150

PALAEOMAGNETISM, Former directions, dips and intensities of terrestrial magnetic fields may be determined by measurement of thermo-remanent magnetism (q.v.) in ancient igneous rocks and so permit plotting the positions in the past of the magnetic poles. Such plots confirm the reality of Continental Drift (q.v.), 104*ff.*, 153

PAUDORF, Lower Austria, The village of this name, near which is a famous exposure of the soil on loess of the Second Interstadial of the Last Glaciation (LGl₂₋₃), has lent its name, more generally, to the interstadial and its deposits elsewhere, 88, 124

PAVEMENT, A wind-swept surface of glacial outwash or desert, from which all finer sediments have been deflated, leaving only coarser blocks and stones exposed, 28

PENULTIMATE GLACIATION (Zeuner), (PGl=Riss/Saale), 76

PENULTIMATE INTERGLACIAL (PIgl), Great Interglacial, Mindel-Riss, Elster-Saale, Holsteinian, Hoxnian etc. Its flora and fauna, 69, 145

PERCHED BLOCKS, Rock-masses gently deposited by melting ice, sometimes in precarious positions, otherwise inexplicable in the case of erratics, 23, Fig. 6

PERIGLACIAL ZONE, Zone beyond the extent of land-ice, but climatically and geologically affected by its presence, 30, 50, 156

PERMAFROST, FROZEN GROUND, TJAELE, Ground permanently frozen in some depth, thawing only superficially in summer, 31

PETROLOGY, Branch of geology concerned with the study of rock-types, their origins, constitutions and classification, 95, 100

PICKERING, Yorks, Ice-dammed lake of Last Glaciation, 89, Fig. 38

PLANKTON, Floating and passively drifting minute vegetable and animal organisms, generally of lakes or seas, 45, 61, 102

PLEISTOCENE (Gk. 'most recent'), The latest geological Period, extending from, perhaps, 3,000,000 up to 10 thousand years B.P., 11, 146

PLUVIAL PERIODS, Phases of wetter climate in the Tropics, corresponding, north of the desert zone, with northern glaciations, south of the deserts, with interglacials, 56

POLLEN-ANALYSIS, Extraction of plant-pollen from peats and soils and sediments, determination of species, counting the grains representing them and plotting the flora for each horizon (spectrum). The changes from horizon to horizon when plotted as a pollen-diagram recapitulate the vegetational history of the site and may indicate a date for an unknown deposit when the pollen-sequence for the area is sufficiently well known and dated, 46, 50, 91

POLYGON-SOILS, Structure-soils in the periglacial zone, bounded in plan by frost-cracks or stone-polygons due to frost-heaving (q.v.), 31ff.

PONDERS END, Lea valley, Essex, Site of an arctic plant-bed attributable to the Second Interstadial of the Last Glaciation (LGl$_{2-3}$). Name applied to this stage in the evolution of the Thames basin, 91

PONGID, An individual or species belonging to the Family Pongidae, great apes, 127

PONTINE MARSHES, Lower Versilia, Italy – sections in important marine and coastal deposits of the Last Glaciation, largely below present sea-level, 87–8

POSTGLACIAL PERIOD or Holocene (completely recent), characterized by the latest (Flandrian) marine transgression and the presence of exclusively modern flora and fauna. Includes the last 10,000 years, 14, 92, 163

POTASSIUM/ARGON (K/A) DATING, a method of isotopic dating applicable to igneous rocks of the older Pleistocene (more than about 250 thousand years old), 62, 95, 145–6

PRECIPITATION, Deposition of moisture on the land-surface, whether by rain, snow, hail or dew, 17

PRIMATE (Zool.), An individual, species or higher zoological group belonging to the Order Primates, 126

PROFILE, RIVER, The longitudinal height/distance graph of a river's course, 35, Figs. 18, 25

PROTOACTINIUM/THORIUM DATING, By measuring the ratio between

Pa²³¹ and Th²³⁰, two radioactive isotopes
with different half-lives, the dates, back to
175 thousand years B.P., of sections of deep-
sea cores may be estimated, 104, Fig. 41, 146
PYROCLASTIC VOLCANIC
ROCKS, Fragmentary materials – bombs,
lapilli, pumice, ash – formed in explosive
eruptions and deposited as airborne
sediments, 99, 100

RADIATION-CURVES, of Milan-
kovitch, Variations of summer solar radiation
received in high latitudes, calculated back for
the last 1 million years, form the basis of a
natural chronology of the later Pleistocene.
Radiation-minima are supposed to have
caused glacial advances, if with some delay
('retardation') difficult to estimate, Fig. 41,
141, 157
RADIOCARBON, C¹⁴, A heavy radioactive isotope of carbon, used to date organic
matter back to about 30,000 years B.P., 91,
104, 144ff.
'RAISED' BEACHES, Marine beach-
platforms, with or without deposits, due to
former seas at levels higher than that of the
present day – see Isostasy, Eustasy, 54
REGRESSION, MARINE, Retreat of
the sea with falling sea-level, baring areas of
continental shelf and locally providing land-
bridges between continents and islands, 35,
Fig. 17, 80, 87, 156, 164
REJUVENATION, of rivers, caused by
eustatic fall of sea-level, or corresponding
tectonic (q.v.) uplift of land, 35
RETARDATION, Time-lag between a
radiation-minimum and the ensuing glaciation
(see RADIATION-CURVES), 142
RETICULATED SOILS, The surface of
frozen ground, marked by net-like (Germ.
Steinnetze) segregations of loose stones. A
feature of periglacial conditions, 31ff., Fig. 32
RISS GLACIATION, The third Alpine
glaciation of Penck & Brückner, Penultimate
Glaciation (PGl), 14, 76ff.

RIVER-AGGRADATION, Climatic,
Fig. 25
    Eustatic, Fig. 18
    DEPOSITS, 46ff.
    EROSION, 47
    PROFILE, 35
    REJUVENATION, 37
ROCK-BASINS, Depressions excavated in
bedrock by moving ice, 23
RÜDERSDORF, Berlin, Last Interglacial
deposits at, 83

SAALE GLACIATION, The second
glaciation of northern Europe hitherto
recognized, =Riss of Alps (PGl), 14, 76ff.
SAND, AEOLIAN, Formation of dunes
by wind-blowing of sand off wide foreshores
laid bare by marine regression; corresponds,
in unglaciated regions, to a glaciation in
higher latitudes, 33
    GRADE, Particles in a sediment 2·0–0·06
    mm. in diameter, 33, 97–8
SANDR (Icelandic), Bare plain of shifting
outwash-deposits in front of an ice-sheet or
glacier, 26
SANGAMON INTERGLACIAL, In
N. America the Interglacial between Illinoian
and Iowan glaciations (LIgl), 83
SCREE, Mass of generally frost-weathered
rock-débris lying at the angle of rest below
steep cliffs, Fig. 4
SEA-LEVEL CHANGES, Eustatic,
isostatic, secular fall (qq.v.), 34ff., 51ff., 95,
147, 156, 164
SEA-LOCHS or FJORDS, Flooded
valleys on coastlines, generally ice-deepened, 23
SECULAR FALL OF SEA-LEVEL,
A steady fall throughout the Pleistocene,
unconnected with the Ice Ages, at a rate
estimated as 12 mm. per century, 95, Fig. 39,
164
SICILIAN SEA-LEVELS, Represented
by several well-marked high beaches and
planations about 100 m. above the present
sea-level, Fig. 39

ICE AGES: THEIR NATURE AND EFFECTS

THERMO-REMANENT MAGNETISM, In ancient igneous rocks a magnetic orientation of the constituent minerals is gained on first solidification and cooling in the contemporary geomagnetic field. This indicates the direction, dip and intensity of that field, which may be determined today from an oriented sample of the rock, 105

TIGLIAN, A Lower Pleistocene (Villafranchian) phase of warm temperate climate. Named from Tegelen, near Limburg, Netherlands, 60

TILL or BOULDER-CLAY, An unstratified, unsorted geological deposit of a glacier or ice-sheet, 12

TILLITE, 'Fossil' boulder-clay, altered to hard stone, of pre-Pleistocene glaciations (e.g. Dwyka Tillite (Permo-Carboniferous), from Nooitgedacht, S. Africa), 21

TJAELE, Scandinavian name for Permafrost or Frozen Ground (qq.v.), 31

TORRE IN PIETRA, Italy, Archaeological site giving Acheulian implements associated with Interglacial animal remains (PIgl), 73

TRAFALGAR SQUARE, London, Gravels of the Last Interglacial with fauna and flora, 123

TRANSGRESSION, MARINE, Flooding by sea of low-lying coasts during rise in sea-level, 149, 164

TYRRHENIAN SEA-LEVEL, Marked by ancient high beaches at about 100 ft. (33 m.) above present, datable to the Great (PIgl) Interglacial, 73

U-SHAPED VALLEY, A valley of characteristic cross-section, certainly once containing a glacier, 20, Fig. 3

UNCONFORMITY, A break or gap in a stratigraphical sequence, generally marked by angular discordance of bedding (e.g. horizontal strata laid down on the eroded edges of older, tilted, beds), 77

UNIFORMITARIANISM, Doctrine of, The principle of Uniform Causes in geology – i.e. that only agencies which can be observed in action at the present day may be adduced to explain phenomena of the geological past, 13

UPPER PALAEOLITHIC, A human cultural stage marked by stone industries based on blades struck from prismatic stone cores. In Europe first introduced in the First Interstadial of the Last Glaciation (LGl_{1-2}), 88, 91

V-SHAPED VALLEY, The usual form, in cross section, of a valley cut by stream-action (as opposed to the U-shape imposed by a valley-glacier), 41, 47

VARVES (Swed.), Annual pairs of sediment-layers (coarse-fine) in lake-sediments transported by streams of glacial melt-water, 40

VEGETATION, Mainly influenced by climatic and soil-conditions, 46, 50

VÉRTESSZÖLLÖS, Hungary, Site of discovery of skull and implements of *H. erectus ungaricus*, dated to the Mindel I–II ($ApGl_{1-2}$) Interstadial, 66, 129

VILLAFRANCHIAN, The lowermost subdivision of the Pleistocene (Villafranca, Italy). Floras and faunas of the same, 57*ff.*, Fig. 28

VOLCANIC DEPOSITS, Mainly pyroclastic (i.e. particulate, rather than solid lavas) in our context, 99, 100

WAALIAN INTERGLACIAL (Netherlands), Sediments attributable to a phase of temperate climate between two early cold periods within the Villafranchian, 60, 62

WARTHE GLACIATION, A glaciation of N. Europe, marked by the Fläming end-moraines, assigned by some to a late stage of Saale (PGl); by Zeuner and others to the first stadium of the Last Glaciation (LGl_1), 14, 76, 86

WEATHERING, Physical and chemical weathering-agencies, 18, 43

WEGENER, ALFRED, Author of Continental Drift Theory, 153
WEICHSEL GLACIATION, Latter phases, at least, of the Last Glaciation of northern Europe (see WARTHE), 14, 86*ff.*
WISCONSIN GLACIATION, In the wider sense, the whole of the Last Glaciation (including the IOWAN) of North America, corresponding to Würm/Warthe + Weichsel in Europe, 14, 86*ff.*
WÜRM GLACIATION, The fourth, and last, Alpine glaciation of Penck & Brückner. Last Glaciation (Zeuner), 14, 86*ff.*

YARMOUTH INTERGLACIAL, In north America the interglacial between Kansan and Illinoian glaciations, equivalent to the Great Interglacial (PIgl) of Europe, 69
YOUNGER DRIFT, Deposits, generally north of a line from the Bristol Channel to the Wash, attributable to the various phases of the Last Glaciation in Britain (Warthe + Weichsel of the Continent), 89

ZEUNER, F. E., Pleistocene geologist and palaeontologist, 15, 76, 86, 141

Bibliography

BERNARD, E. A., 'Interprétation astronomique des pluviaux et interpluviaux du Quaternaire africain.' Actes du IVe Congrès panafricain de Préhistoire et de l'étude du Quaternaire (1962), 67–95.
BLANC, A. C., 'Giacimento ad industria del Paleolitico inferiore (Abbevilliano ed Acheuleano) e fauna fossile ad *Elephas*, a Torre in Pietra, presso Roma', Rivista di Antropologia (1954), *41*.
BREUIL, H. and KOSLOWSKI, 'Études de stratigraphie paléolithique dans le nord de la France, la Belgique et l'Angleterre', l'Anthropologie, *41* (1931), 449.
BUCKLAND, W., *Reliquiae Diluvianae*, London (1823), John Murray.

CAILLEUX, A., 'Les actions éoliennes périglaciaires en Europe', Mém. Soc. géol. de France, *21* (146) (1942), 176 pp.
CLAYTON, K. M., 'Some aspects of the glacial geology of Essex', P.G.A., *68* (1957), 1–21.
CORNWALL, I. W., 'Outline of a stratigraphical "bridge" between the Mexico and Puebla Basins', Univ. of London, Inst. of Archaeol. Bulletin No. 7 (1968), 89–140. Ibid (1969), No. 8.

DURY, G. H., *The face of the Earth*, Pelican Books (1959), A.447, Harmondsworth, 223 pp.

EBERL, B., 'Die Eiszeitenfolge im nördlichen Alpenvorlande', Augsburg (1930), 427 pp.
EMILIANI, C., 'Pleistocene temperatures', J. Geol., *63* (1955), (6), 538–78.
EMILIANI, C., *The significance of deep-sea cores*, Brothwell & Higgs (Eds.), Science and Archaeology (1964), 99–107, Thames & Hudson, London.
ERICSON, D. B. and WOLLIN, G., (U.S. publ. 1964) *The deep and the past*, London (1966), Jonathan Cape. 292 pp.
EVERNDEN, J. F. and CURTIS, G. H., 'Age of Bed I, Olduvai Gorge, Tanganyika', Nature *191*, No. 4787 (1961), 478–9.
EWING, M. and DONN, W. L., 'A theory of ice-ages', Science, *123* (1956, 1958), 1061–6; *127*, 1159–62.

FAIRBRIDGE, R. W., 'The changing level of the sea', Sc. Amer., *202* (1960), 70–79.
FLINT, R. F., 'Glacial and Pleistocene geology', New York, J. Wiley (1957).
FRECHEN, J. 'Die Herkunft der spätglazialen Bimsstuffe in mittel and süddeutschen Mooren', Geol. Jahrb., *67* (1952), 209–30.

GALLANT, R. C., *Bombarded Earth*, London (1964), John Baker, 256 pp.

GEIKIE, J., *The Great Ice Age*, London (1894), Stanford, 850 pp.

GENTNER, W. and LIPPOLT, H. J., 'Age of the basalt-flow at Olduvai, East Africa', Nature, *192* (1961), 720–1.

HAUG, E., Traité de Géologie, 1760–1921, Paris (1911).

HOLMES, A., *Principles of physical geology*. Nelson, London (1965), 1288 pp.

KING, W. B. R. and OAKLEY, K. P., 'The Pleistocene succession in the lower parts of the Thames valley', Proc. prehist. Soc., N.S.*2* (1936), 52–76, London.

LYELL, C., *Principles of Geology*, London (1830–33), J. Murray.

MACDOUGALL, I. and CHAMALAUN, F. H., 'Geomagnetic polarity-scale and time,' Nature, *212* (1966), 1415.

PENCK, A. and BRÜCKNER, E., 'Die Alpen im Eiszeitalter', Leipzig (1901–9), 1189 pp.

PROŠEK, F. and LOŽEK, Strati-graphische Übersicht des tschechoslowakischen Quartärs, Eiszeitalter und Gegenwart, *8* (1957), 37–90.

ROSHOLT, J. N., et al., 'Absolute dating of deep-sea cores by the Pa^{231}/Th^{230} method', J. of Geol., *69* (1961), 162–85.

SOERGEL, W., 'Die Gliederung und absolute Zeitrechnung des Eiszeitalters', Fortschr. Geol. Palaeont., Berlin, *13* (1925), 125–251.

STOKES, 'Another look at the Ice Age', Science, *122* (1955), 815–21.

WEGENER, A., *The origin of continents and oceans*, 3rd ed., Methuen, London (1924), 212 pp.

WEST, R. G., 'Problems of the British Quaternary', Proc. Geol. Assoc., *74* (1963), 127–86.

WEST, R. G., *Pleistocene Geology and Biology*, Longmans, London (1968), 377 pp.

WOLDSTEDT, P., *Das Eiszeitalter*, Stuttgart (1954–58).

WOOLDRIDGE, S. W., 'The Pliocene history of the London Basin', Proc. Geol. Ass., *38* (1927), 49.

WOOLDRIDGE, S. W., and CORNWALL, I. W., 'A contribution to a new datum for the prehistory of the Thames valley', Univ. Lond. Inst. Archaeol. Bulletin No. 4 (1964), 223–32.

ZEUNER, F. E., *The Pleistocene Period*, London (1945) (2nd ed. 1959), Hutchinson.

ZEUNER, F. E., 'Soils and shorelines as aids to chronology', U.L. Inst. Arch. Bull. No. 4 (1964), 233–50.

Textbooks for further reading

CHARLESWORTH, J. K., *The Quaternary Era*. 2 vols, Arnold, London (1957), 1700 pp.

EMBLETON, C. and KING, C. A. M., *Glacial and periglacial Geomorphology*. Arnold, London (1968), 608 pp.

FLINT, R. F., *Glacial geology and the Pleistocene epoch*. Wiley, New York (1947), 589 pp.

HOLMES, A., *Principles of physical geology*. Nelson, London (1965), 1288 pp.

WEST, R. G., *Pleistocene geology and biology*. Longmans, London (1968), 377 pp.

WOLDSTEDT, P., *Das Eiszeitalter*. Stüttgart (1954–58).

ZEUNER, F. E., *The Pleistocene Period*. Hutchinson, London (1945), (2nd ed. 1959), 447 pp.